D0746413

MAPPING DESIGN RESEARCH

Simon Grand, Wolfgang Jonas (Eds.)

MAPPING DESIGN RESEARCH

Birkhäuser
Basel

CONTENTS

Essay

FOREWORD BIRD

Design research is becoming increasingly important in academic and creative contexts within the international research landscape. There is a special interest in research approaches that take their cue from design practices, processes and issues and attempt to closely link knowledge production to the world of design. Key topics include the researchers' involvement in their own experimental systems and the complex interplay between writing systems, data generation, and data interpretation. Nevertheless, it often remains unclear which concepts, ideas, and stereotypes concerning design and research form the basis of so-called practice-led design research. At times the rather paradoxical impression may arise that the localization of independent research focuses within the field of design not only draws on traditional models of research, science and knowledge production, but also attempts to question and further develop such models.

Against this backdrop, the present book endeavors to define the foundations of the design research that is taking place in the present and, more importantly, that will take place in the future. The study takes the form of a topological overview, but is also informed by a dialogical structure: groundbreaking contemporary concepts and approaches that define design as a research activity and characterize research as a practical undertaking are identified and set in relation to each other. The purpose is not to demonstrate the existence of an unbridgeable divide between various knowledge cultures or to emphasize their antithetical features. Rather, through its texts and introductory essays, this book aims to help map out the productive interactions, parallels, and overlaps between those concepts in design and research that focus on the common theme of design and knowledge production. The selected perspective illustrates that, although there are distinctions between academic and design research, they are both constituted by specific epistemic, aesthetic, and social practices.

The importance of the focus on pivotal texts for current and future design research cannot be overstated. However, while certain charismatic names come up again and again in the current discussion, the community has thus far lacked a handbook that is well-founded and conscientiously selected while also being straightforward and accessible. Grand and Jonas have now filled this gap. Compiled on the basis of a deep and extensive knowledge of the field, their collection of some eighteen international texts from the period after 1960 mirrors the principal developmental phases of design research in particular and in the relevant fields of scientific research.

Board of International Research in Design, BIRD

Simon Grand, Wolfgang Jonas

INTRODUCTION

Basic Idea

The issue of foundations for design research is controversial, at least when researchers claim to follow their own "designerly" paradigm of knowledge generation. The controversy arises because "designerly" inquiry deviates from the taken-for-granted epistemological traditions, methodological guidelines, and academic standards set by the more established disciplines and faithfully followed and enforced by many in Design. Simply put, there is the "scientific" path and then there is the attempt to develop a "designerly" theory of knowledge production, the latter being associated with notions like "practice-led research", "project-grounded research", and "research through design", all of which have generated considerable interest in recent years.

The "scientific" approach tries to adopt seemingly rigid scientific standards without reflecting on their appropriateness for design research. External standards are transferred wholesale into design; a "colonization" of the design field, one might say. We argue strongly for the "designerly" approach instead. The central question is: how to substantiate this claim in the face of attempts to make design research a strictly scientific endeavor? More explicitly: *How can design establish its own genuine research paradigm, independent of the sciences, the humanities, and the arts, one that is appropriate for dealing with purposeful change in complex, ill-defined, real-world situations?*

Our starting point is the observation that those established and respected disciplines are themselves undergoing changes. Non-traditional epistemological approaches are gaining in importance. The relationship of research to practice is changing. Empirical studies reveal that the processes of the Natural and Engineering Sciences are design-like. The relationship between rigor and relevance in inquiry is increasingly delicate. Furthermore, we argue that a new and strong process of convergence between the Sciences and Design is happening and this convergence is heading in the direction of a new "trans-domain". In our opinion, Design risks being caught in the past if it misses these developments.

The academic cultures of the sciences and the arts were still very much integrated during the Renaissance. Their separation from the 17th century onward finally led to what we know as "traditional science" and "traditional design" today. Supposedly, Science produces theoretical knowledge, which Design applies in practice. Despite the various arguments advanced to challenge this black-and-white dualism, and various past efforts from Design to overcome it, today's mainstream design discourse still rests on it. The Bauhaus, the New Bauhaus, the Ulm school—and even the Design Methods Movement to some extent—have attempted to reconcile the split, but their influence on research practice have been limited. Now, however, we observe strong new signs of convergence from the Sciences and Design. On the one hand, the Sciences have begun to reflect on the social-embeddedness and the context-dependency of their knowledge production. On the other hand, Design has gradually become aware of the intensity of knowledge generation in their working processes and has begun to develop their own notions, methods, and standards of research.

We take the early designerly attempts seriously and develop them further. Instead of adapting to nineteenth century scientific standards, we take a closer look at recent inquiries into the nature of scientific research practices and compare them to inquiries into the nature of design and design research. Strong similarities between scientific and designerly processes are showing up. So, instead of establishing the clear split, which implies deficiency in the designerly research, we demonstrate a convergence of scientific and designerly approaches, with clear distinctions regarding the character of the outcomes, of course. Or, in other words, we create a narrative that sets out to reveal the common designerly core of both approaches to knowledge generation. As a projection, which is not further substantiated here, we put forward the hypothesis of a convergence towards a new "trans-domain". The book consists of two essays and two collections of texts.

Part I: Exploring the Swampy Ground: The Logic of Design Research

The first essay, titled "Exploring the Swampy Ground—an inquiry into the logic of design research", describes the development of design research concepts, with special emphasis on the above-mentioned "designerly" concepts. It is followed by a collection of core / paradigmatic texts that are useful to substantiate this claim:

1) Ranulph Glanville (1980) is the seminal text in which design thinking is presented as the generic form of knowledge production. The main argument is that scientific research is a specialized and highly formalized sub-discipline of design research.

2) Warren Weaver (1948) introduces the notion of "organized complexity", crucial for conceptualizing design thinking and design research. His programmatic considerations in terms of dealing with organized complexity can be regarded as ideas about transdisciplinarity and mode 2 science *avant la lettre*.

3) Herbert Simon's text, first published in the 2nd edition of his *The Sciences of the Artificial* (1981), shows an often neglected aspect of his thinking. The non-causal and evolutionary nature of social designing is depicted as one of its specific traits. He further argues that "Design, like science, is a tool for understanding as well as for acting".

4) Charles Owen (1998) presents the design research approach of the Institute of Design in Chicago. At the same time, it is a strong plea in favor of following a designerly way of knowledge production and

against the "colonization" of design research by external standards and processes of inquiry.

5) Christopher Frayling (1993), often cited and rarely read, introduces the controversial distinctions between research *into*, *for*, and *through* art and design. We consider this a useful starting point for the development of a designerly paradigm of design research.

6) Bruce Archer (1995), a kind of "late work", presents a clear and distinct summary of his position regarding design research. He uses the categorization of research about, for, and through design, and offers a consideration of the latter as the most interesting and most challenging one.

7) Alain Findeli (2008) is his precise attempt at clarifying the difference between scientific and designerly research and the epistemological characteristic of "project-grounded research".

8) Ken Friedman (2003) tries to introduce the scientific categories of basic, applied, and clinical research into design research, and can be considered a manifesto for the adoption of classical scientific standards into design research.

9) John Christopher Jones (undated) presents a deeply philosophical, poetic, and in a way timeless, account of practice, theory, and research in design.

Part II: Research as Design: Promising Strategies and Future Possibilities

The second essay, entitled "Research as Design: Promising Strategies and Possible Futures", reflects design research in the light of science studies. First, the essay discusses central qualities of *design as practice*, referring to the discussion of "designerly" ways of knowing (see the first essay). Second, the essay introduces major insights from science studies, which are conceptualizing and empirically describing practices and processes of scientific research; major parallels between research processes and design processes are explored (*research as design*) as a basis for developing new perspectives on design research. Third, the essay looks at Design Fiction and Critical Design as two current research programs that relate design research to science studies, and thus make it possible to identify promising strategies for design research, assembled in a *Design Fiction Method Toolbox*.

This second essay is followed by excerpts from book chapters and scientific articles that discuss different aspects of scientific research. We differentiate three sections, assembling three papers for each:

A. Scientific research as a controversial process of knowledge production

1) Peter Galison (1997) explains why the disunification of science is not a problem, but actually an important premise for scientific advancement and the strength of scientific research. It is the continuous scientific controversies and debates, as well as the interactions and exchanges between disciplines, methods, instruments, and theories in so-called "trading zones" that characterize the successful practice of scientific research. In this line of argument, design research can be interpreted as entering these trading zones.

2) Michael Gibbons, Camille Limoges, Helga Nowotny, Simon Schwartzman, Peter Scott, and Martin Trow (1994) describe how scientific research and its predominant mode of knowledge production are shifting from a disciplinary mode 1 approach towards a transdisciplinary mode 2 approach. Mode 2 not only transcends disciplinary boundaries, but also emphasizes the societal, cultural, and political embeddedness of research. It is productive to interpret design research as a particular field of mode 2 research.

3) In the tradition of the sociology of translation, Michel Callon (1999) describes the processes of scientific research as translation, arguing that through the problematization of established knowledge and perspectives, the interessement of heterogeneous parties, the enrolement of relevant actors, and the mobilization of multiple resources, taken-for-granted, actual scientific knowledge is questioned, discussed, and transformed into new knowledge. Design practice and design research are involved and engaged in such translation processes.

B. Scientific research as experimentation within and beyond the laboratory

4) Karin Knorr Cetina (1999) describes the particular qualities of laboratories as the paradigmatic locus of scientific knowledge production, emphasizing the inherently situational, contextual, material, visual, textual, and processual qualities of scientific research. Laboratories can be seen as material, intellectual, and symbolic infrastructures that make experimentation possible as an exemplary way of conducting scientific research. The importance of experimentation in design research provides a good basis for relating it to the laboratory studies.

5) Hans-Jörg Rheinberger (1997) describes experimental systems as the "smallest integral working units of research". They are characterized as systems enabling scientific experimentation, by providing the methods, tools, instruments, technologies, and materials needed to ask and to answer relevant new research questions in the experimentation process. Experimental systems can thus be characterized as "machines for making the future". The notion of the experimental system is productive for distinguishing design research from design practice.

6) Bruno Latour (2004) argues that important contemporary experiments are no longer taking place inside a laboratory with well-defined boundaries. Instead, we now observe collective experiments and related controversies, which explore and research, and interpret and explain, globally relevant transformation processes. From ubiquitous computing to the financial crisis, and from vaccination initiatives to global warming, we are all involved in such collective experiments. Design as practice and research is engaged in such experiments in multiple ways.

C. Messiness in scientific research and future perspectives

7) John Law (2004) describes scientific research as a messy process, on the one hand constantly challenging and questioning, and criticizing and subverting, existing knowledge and taken-for-granted descriptions and explanations of the world as it is, while on the other hand creating and establishing new knowledge, descriptions, and explanations, which are again assembled and packed into taken-for-granted, unquestioned, self-evident new perspectives. This provides a productive starting point for exploring and structuring the practice of design research.

8) Julian Bleecker (2009) explicitly relates insights from science studies to design practice in order to establish a research program that he calls Design Fiction: thinking through fiction to invigorate what design could be, as well as what new worlds could be created, beyond the routine, everyday notion of what design does, and beyond the world as it is. What if? is a central question in Design Fiction, and a leading question in design research, when focusing more on future possibilities than on taken-for-granted realities.

9) Anthony Dunne (2005) complements this approach with his research program of Critical Design, an explicitly critical and subversive approach. In this perspective, speculative design and prototypical artifacts are used to challenge narrow premises, assumptions, preconceptions, and givens about existing artifacts and technologies, and thus to raise awareness, expose assumptions, provoke action, spark debate, and even entertain, as ways of getting researchers and designers to engage with Critical Design.

Based on these insights, a bundle of research strategies and design methods are deduced and assembled into a *Design Fiction Method Toolbox*, as a way of translating the fundamental debates, major perspectives, and multiple approaches in science studies into a set of research practices that can be directly related to the particular qualities and designerly ways of knowing that characterize design practice and design research. As a projection, we take the hypothesis of a convergence towards a new "trans-domain" of Science and Design as a promising starting point for exploring new strategies and possible futures for the field of Design Research. Opening up for new possibilities and opportunities, rather than closing down on defined positions and perspectives, is the guiding spirit of this book.

We wish to thank the staff at Birkhäuser and, in particular, Dr. Robert Steiger for their endless patience and support in making this book a reality. Special heartfelt thanks go to our colleague Ralf Michel, who played a key role in developing, planning, and producing this publication as one of its initiators.

YBP Library Services

MAPPING DESIGN RESEARCH; ED. BY SIMON GRAND.

Paper 263 P.

BASEL: BIRKHAUSER VA, 2012

NEW PAPERS ON PRODUCTIVE INTERACTIONS, PARALLELS, AND OVERLAPS BETWEEN CONCEPTS IN DESIGN & RESEARCH

ISBN 3034607164 Library PO# SLIP ORDERS

List 100.00 USD

6207 UNIV OF TEXAS/SAN ANTONIO Disc 17.0%

App. Date 4/24/13 MGT.APR 6108-09 Net 83.00 USD

SUBJ: INDUSTRIAL DESIGN.

CLASS TS171 DEWEY# 745.2 LEVEL ADV-AC

YBP Library Services

MAPPING DESIGN RESEARCH; ED. BY SIMON GRAND.

Paper 263 P.

BASEL: BIRKHAUSER VA, 2012

NEW PAPERS ON PRODUCTIVE INTERACTIONS, PARALLELS, AND OVERLAPS BETWEEN CONCEPTS IN DESIGN & RESEARCH

ISBN 3034607164 Library PO# SLIP ORDERS

List 100.00 USD

6207 UNIV OF TEXAS/SAN ANTONIO Disc 17.0%

App. Date 4/24/13 MGT.APR 6108-09 Net 83.00 USD

SUBJ: INDUSTRIAL DESIGN.

CLASS TS171 DEWEY# 745.2 LEVEL ADV-AC

Wolfgang Jonas

EXPLORING THE SWAMPY GROUND

An inquiry into the logic of design research

1 Introduction, assumptions, and purpose

Prologue

> *"Now! Now!" cried the Queen. "Faster! Faster!" And they went so fast that at last they seemed to skim through the air, hardly touching the ground with their feet, till suddenly, just as Alice was getting quite exhausted, they stopped, and she found herself sitting on the ground, breathless and giddy.*
>
> *The Queen propped her up against a tree, and said kindly, "You may rest a little, now."*
>
> *Alice looked round her in great surprise. "Why, I do believe we've been under this tree the whole time! Everything's just as it was!"*
>
> *"Of course it is," said the Queen. "What would you have it?"*
>
> *"Well, in our country," said Alice, still panting a little, "you'd generally get somewhere else—if you ran very fast for a long time as we've been doing."*
>
> *"A slow sort of country!" said the Queen. "Now, here, you see, it takes all the running you can do, to keep in the same place. If you want to get somewhere else, you must run at least twice as fast as that!*
>
> *(Carroll 1996, 151–152)*

This essay is an attempt at a clustering logic for design research. *Scientific research* is about generating new knowledge in distinct fields and thus constructing ever more elaborate disciplinary terminologies and theories, which allow explanation and prediction relating to natural and even social phenomena. This knowledge can be applied to dealing with, not necessarily solving, real-world problems.

Figure 1
The paradox of the Red Queen, or, progress in design research?

Knowledge has different meaning, status, and use in science and in design. *Science* is aiming at predictability, and thus needs stable models that repeatedly deliver "the same." Science has to purify its models in order to convert them from vague hypotheses into prediction machines. Bodies, consciousnesses, communications, and artifacts are neatly separated. Scientific problems are solved, as long as the solution does not turn out to be false, "false" in this context meaning of less explanatory power than a new solution. *Design,* on the other hand, is aiming for single new phenomena that must be able to fit various unforeseeable conditions. Design has to intentionally create variations—differences—because the "fits" will dissolve, fade away, grow old-fashioned. Design environments change too fast to be able to speak of "true" or "false" design knowledge. The archive of design knowledge is like a memory, a growing reservoir of variations as well as restrictions. Design expertise seems to be the art of dealing with scientific and non-scientific knowledge, with fuzzy and outdated knowledge, and with no knowledge at all, in order to achieve these value-laden fits. The "art of muddling through," or, more positively, of "informed intuition," should not be scorned, but seen as a core element of designing.

We are facing the paradoxical situation of increasing manipulative power through science and technology and, at the same time, decreasing prognostic control of its social, economic, and ecologic consequences. As soon as a solution is found, it becomes the nucleus of a new problem. The emerging discourse of *mode 2 science,* with its primacy of application contexts, transdisciplinarity, project-orientation, social accountability, and new quality criteria, is a response to these new conditions. *Contextualized scientific problems are design problems*. The new scientific criterion of social robustness (Nowotny et al. 2001) requires uninterrupted feedback (and feedforward), with its context in the agora. *Mode 2 science is design-like.*

Accepting the limits of project-oriented science and acknowledging its similarity with design suggests a new role for design: at once more modest and more self-assured. More modest in its claim to solve problems, and more self-assured in its claim to present its own designerly paradigm of project-based knowledge production. A new conception of design thinking and design research is required, one which is adequate to this integrating function. We may return to Glanville (1980 → Text 1), who, in his far–seeing classical paper, conceived *research as a design activity* and viewed scientific research as a restricted subdiscipline of design research:

... Under these circumstances, the beautiful activity that is science will no longer be seen as mechanistic, except in retrospect. It will truly be understood honestly, as a great creative and social design activity, one of the true social arts. And its paradigm will be recognized as being design.

Thus design will take its true place as the basis for the activities that create scientific (as well as other) knowledge, and will no longer be sneeringly and trivially dismissed by those who adjudicate without creating, and who are fooled into believing that science is as she is writ. There will be no need for a special area of design research, for all research will be seen to be part of design research, with that which we call, now, design research being the most basic of all.

We are not the first to accept the challenge of mapping or even defining design research. Every attempt makes certain initial distinctions according to its particular understanding of design research. Different categorizations follow from the respective initial distinctions, and we are aware of this unavoidable blind spot. The aim here is different from Bayazit's (2004) historical overview, *Forty Years of Design Research*. Ours is a contribution to the *design of design research*, a design proposal, which is meant to provide a framework for the collection of texts that follow it. We see an evolution of schemes since the 1960s that account for the specific "designerly" ways of producing knowledge (part 2). Despite the ongoing attempt at "scientizing" design research, there are more recent attempts to take up and develop the original approaches (part 3). We try to synthesize this and present the concept of a new scheme (part 4).

2 Attempts to define and map design research

Weaver (1948 → Text 2) paved the way for design research by introducing the concept of "organized complexity," a central category for describing and dealing with design problems. His systemic, normative, and transdisciplinary approach seems to anticipate the mode 2 concept of science. Simon (1969) was one of the first to conceive design, or the making of the artificial, as a distinct *subject matter* and *form* of research, different from the Sciences and the Humanities. He argues that design research is not conducted for its own sake, but in order to improve real-world situations, to "transfer existing situations into preferred ones." The concept of *relevance* shows up here, a concept that seems to be in permanent conflict with the scientific requirement of *rigor*. This polarity may ultimately dissolve in a pragmatist view.

In positioning design at the *interface* between the artifact and its contexts, Simon introduced the idea of *situatedness* and context-dependency in design research. The parallels to the complex of problems that the debate of mode 2 science raised 20 years later are obvious (Simon 1969):

An artifact can be thought of as a meeting point—an "interface" in today's terms—between an "inner" environment, the substance and organization of the artifact itself, and an "outer" environment, the surroundings in which it operates. If the inner environment is appropriate to the outer environment, or vice versa, the artifact will serve its intended purpose. . . .

Alexander puts it similarly (1964, 15):

. . . every design problem begins with an effort to achieve fitness between two entities: the form in question and its context. The form is the solution to the problem; the context defines the problem. In other words, when we speak of design, the real object of discussion is not the form alone, but the ensemble comprising the form and its context. Good fit is a desired property of this ensemble which relates to some particular division of the ensemble into form and context. . . . we may even speak of culture itself as an ensemble in which the various fashions and artifacts which develop are slowly fitted to the rest.

Simon (1996 → Text 3) also pointed to the *evolutionary* and iterative character of design (1996, 163):

The idea of final goals is inconsistent with our limited ability to foretell or determine the future. The real result of our actions is to establish initial conditions for the next succeeding stage of action. What we call 'final' goals are in fact criteria for choosing the initial conditions that we will leave to our successors.

How do we want to leave the world to the next generation? What are good initial conditions for them? One desideratum would be a world offering as many alternatives as possible to future decision makers, avoiding irreversible commitments that they cannot undo. . . .

In the first issue of *Design Studies*, Archer (1969) introduces "Design as a discipline." It is obvious that design research must include a huge diversity of disciplines in order to become productive at all. The prolific paradox of the "undisciplined!" discipline (2008) has been present from the very beginning. Archer (1981, 30) took a Wittgensteinian stance, when he said *"that my own approach to finding an answer to the question* What is Design Research? *is to try to discover what design researchers actually do."* He gives the following definition:

Design Research . . . is systematic enquiry whose goal is knowledge of, or in, the embodiment of configuration, composition, structure, purpose, value and meaning in man—made things and systems .

which is very similar to Findeli's (2008b) recent definition:

> *Design research is a systematic search for and acquisition of knowledge related to general human ecology considered from a "designerly way of thinking" (i.e., project-oriented) perspective.*

Archer lists 10 areas of Design Research, from which "constituent sub-disciplines" emerge, namely Design Phenomenology, Design Praxeology, and Design Philosophy:

Cross follows this scheme and suggests that (1999, 6): "... *design research would therefore fall into three main categories, based on people, process, and products"* and even relates them to historical epochs, the 1920s, the 1960s, and the 2000s (2001). Findeli's (2005) separation into *esthetics*, *logic,* and *ethics* seems to be comparable.

Table 1
Archer's 10 areas and 3 emergent sub-disciplines of Design Research (1981: 31, 35).

Design Phenomenology	
1. Design history	The study of what is the case, and how things came to be the way they are, in the Design area.
2. Design taxonomy	The study of the classification of phenomena in the Design area.
3. Design technology	The study of the principles underlying the operations of the things and systems comprising designs.
Design Praxeology	
4. Design praxeology	The study of the nature of design activity, its organization and its apparatus.
5. Design modeling	The study of the human capacity for the cognitive modeling, externalization, and communication of design ideas.
6. Design metrology	The study of measurement in relation to design phenomena, with special emphasis on the handling of non-quantitative data.
Design Philosophy	
7. Design axiology	The study of worth in the Design area, with special regard to the relationships between technical, economic, moral, social, and esthetic values.
8. Design philosophy	The study of the logic of discourse on matters of concern in the Design area.
9. Design epistemology	The study of the nature and validity of ways of knowing, believing, and feeling in the Design area.
10. Design pedagogy	The study of the principles and practice of education in matters of concern to the Design area.

Table 2
Cross's 3 main categories of Design research (1999, 2001) and—in parentheses—Findeli's (2005) categories.

People (ethics)	Design epistemology – study of designerly ways of knowing; the 2000s
Process (logic)	Design praxeology – study of the practices and processes of design; the 1960s
Products (esthetics)	Design phenomenology – study of the form and configuration of artefacts; the 1920s

Cross emphasizes (2001, 7):

> *... that we do not have to turn design into an imitation of science, nor do we have to treat design as a mysterious, ineffable art. (...) we must avoid totally swamping our research with different cultures imported either from science or art....*

He calls his scheme "designerly ways of knowing," claiming that design is a genuine way of producing knowledge, different from science and art, which, though fuzzier, is very much in accordance with Simon (1969).

In the 1920s, with the modern desire to "scientize" design and the longing for objectivity and rationality, Cross diagnoses the search for scientific design products (e.g., Le Corbusier's "machine for living").

In the 1960s, named the "design science decade" by Buckminster Fuller (1999), there is the search for the scientific design process. The Conference on Design Methods (London, September 1962) marked the beginning of the Design Methods Movement, with its desire to base the process on objectivity and rationality. The "Sciences of the Artificial" (Simon 1969) designate the culmination and the watershed of this development. Simon himself, in chapter 6 on "Social Planning: Designing the Evolving Artifact," made a considerable shift in acknowledging the evolutionary character of every design process, determined by unforeseeable human and social aspects.

In pointedly illustrating the fundamental paradoxes that occur when design (as an activity projecting what should be) is considered as a scientific endeavor (analyzing what is), Rittel (1972) made contributions to this debate that can hardly be overstated. The theory backlash of the 1970s obstructed the growth of these still-vague ideas and it took a decade to recover. Recently, Krippendorff (2007) has sharpened the argument further and called Design Research an "oxymoron," a contradiction in itself. Cross sums up (1999, 51):

> *... The Design Research Society's 1980 conference on "Design. Science: Method" provided an opportunity to air many of these considerations. The general feeling from that conference was, perhaps, that it was time to move on from making simplistic comparisons and distinctions between science and design; that perhaps there was not so much to learn from science after all, and that perhaps science rather had something to learn from design....*

In the 2000s, Cross detects the re-emergence of design-science concerns with a focus on people. This phase of the "products-process-people" model shows a striking parallel to what Findeli later presents as the "Bremen model" (2005), wherein he describes a shift of concern in design research from esthetics (products) to logic (process) and finally towards ethics (people), both on the production side ("upstream") and on the reception side ("downstream") of the design cycle. Cross (2001) tries to clarify the confusion about the relationship between design and science:

Table 3
Interpretations regarding the relationship between Design and Science (Cross 2001).

Scientific Design Design with scientific and other foundations	(52): "So we might agree that scientific design refers to modern, industrialized design—as distinct from pre-industrial, craft-oriented design—based on scientific knowledge but utilizing a mix of both intuitive and non-intuitive design methods. 'Scientific design' is probably not a controversial concept, but merely a reflection of the reality of modern design practice."
Design Science Design as Science	(53): "So we might conclude that design science refers to an explicitly organized, rational, and wholly systematic approach to design; not just the utilization of scientific knowledge of artifacts, but design in some sense as a scientific activity itself. This certainly is a controversial concept, challenged by many designers and design theorists. . . ."
Science of Design Design as a subject matter of Science	W. Gasparski and A. Strzalecki ("Contributions to Design Science: Praxeological Perspective" in *Design Methods and Theories* 24: 2 [1990]): "The science of design (should be) understood, just like the science of science, as a federation of subdisciplines having design as the subject of their cognitive interests." (Cross 53): "In this latter view, therefore, the science of design is the study of design—something similar to what I have elsewhere defined as 'design methodology'; the study of the principles, practices, and procedures of design. . . . The study of design leaves open the interpretation of the nature of design. So let me suggest here that the science of design refers to that body of work which attempts to improve our understanding of design through 'scientific' (i.e., systematic, reliable) methods of investigation. And let us be clear that a 'science of design' is not the same as a 'design science'."
Design as a Discipline	(54): "Despite the positivist, technical-rationality basis of The Sciences of the Artificial, Simon did propose that 'the science of design' could form a fundamental, common ground of intellectual endeavor and communication across the arts, sciences, and technology. What he suggested was that the study of design could be an interdisciplinary study accessible to all those involved in the creative activity of making the artificial world. . . . Design as a discipline, therefore, can mean design studied on its own terms, and within its own rigorous culture. It can mean a science of design based on the reflective practice of design: design as a discipline, but not design as a science. . . . The underlying axiom of this discipline is that there are forms of knowledge special to the awareness and ability of a designer, independent of the different professional domains of design practice. . . . we must avoid swamping our design research with different cultures imported either from the sciences or the arts. . . ."

Cross worries about the "swamping" of design research, but we are in the swamp already, necessarily. The community owes the beautiful metaphor of the "swampy lowland" to Schön. He challenges the Design Science Movement and argues for an epistemology of practice that is implicit in the artistic, intuitive processes, instead. "We have never been modern." (Latour 1998). In his *Reflective Practitioner* (1983), Schön explicitly raises the issue of rigor vs. relevance (42, 43):

> *The dilemma of "rigor or relevance" arises more acutely in some areas of practice than in others. In the varied topography of professional practice, there is a high, hard ground where practitioners can make*

effective use of research-based theory and technique, and there is a swampy lowland where situations are confusing "messes" incapable of technical solutions. The difficulty is that the problems of the high ground, however great their technical interest, are relatively unimportant to clients or to the larger society, while in the swamp are the problems of greatest human concern. . . .

There are those who choose the swampy lowlands. They deliberately involve themselves in messy but crucially important problems and, when asked to describe their methods of inquiry, they speak of experience, trial and error, intuition, and muddling through.

Other professionals opt for the high ground. Hungry for technical rigor, devoted to an image of solid professional competence, or fearful of entering a world in which they feel they do not know what they are doing, they choose to confine themselves to narrowly technical practice.

Owen (1998 → Text 4) can be positioned in the pragmatist tradition, too, even if his approaches appear more "scientific" than Schön's. Owen is one of those who believe that design is a special form of knowledge production and that, although design's own research culture is still young and weak, the importation of methods from more established disciplines does not necessarily contribute to the development of the discipline (1998, 10):

those who seek to work more rigorously look to scientific and scholarly models for guidance, and we find references to "design science" and examples of "design research" that would seem to fit more appropriately in other fields.

Yet, it is reasonable to think that there are areas of knowledge and ways of proceeding that are very special to design, and it seems sensible that there should be ways of building knowledge that are especially suited to the way design is studied and practiced. . . .

Owen analyzes the knowledge-building and knowledge-using processes in various scientific and non-scientific disciplines and arrives at the conclusion that these processes are fundamentally the same for inquiry and application. The differences lie mainly in the purposes of the activity, and in the codes and value bases that are used.

In order to illustrate the graduate studies at the Institute of Design in Chicago, Owen presents a clustering according to the well-known analytic-synthetic / symbolic-real matrix, which is used at ID Chicago as a universal structural framework and which can be traced back to Kolb's pragmatist learning cycle (1984) and even earlier models. Analytic is related to

Table 4
Differences in measures and values (Owen 1998).

Domain	Discipline	Measures	Source of Values
Science	Mathematics	True / false Correct / incorrect Complete / incomplete	Reason Logic
	Chemistry	True / false Correct / incorrect Right / wrong Works / doesn't work	Physical world
Technology	Mechanical engineering	Right / wrong Better / worse Works / doesn't work	Physical world Artificial world
Law	Statutory law	Just / unjust Lawful / unlawful Right / wrong	Social contract
Arts	Painting	Beautiful / ugly Skillful / unskilled Thought-provoking / banal	Culture
Design	Product design	Better / worse Beautiful / ugly Fits / doesn't fit Works / doesn't work	Culture Artificial world

research, synthetic to professional practice, symbolic to planning, real to human-centered design (17, 18):

> *"Design Planning Research students investigate and develop theory, methods and processes for planning and concept formation. Human-Centered Research students investigate and develop theory, methods and process for the detailed design of systems and services and their incorporated products and communications. Design Planning Professional students apply the tools of design planning to the creation of design plans for institutions and industry. Human-Centered Design Professional students apply the tools of human-centered design to problems of systems and services with their associated products and communications."*

Obviously Owen concentrates on building knowledge FOR the improvement of the design / planning process (left side) and on applying this knowledge in design / planning (right side). The feedback loops indicate that the knowledge base is supported THROUGH the design / planning processes. This is the pragmatist focus, which integrates inquiry and application. Finally, Owen gives a number of recommendations, such as (19):

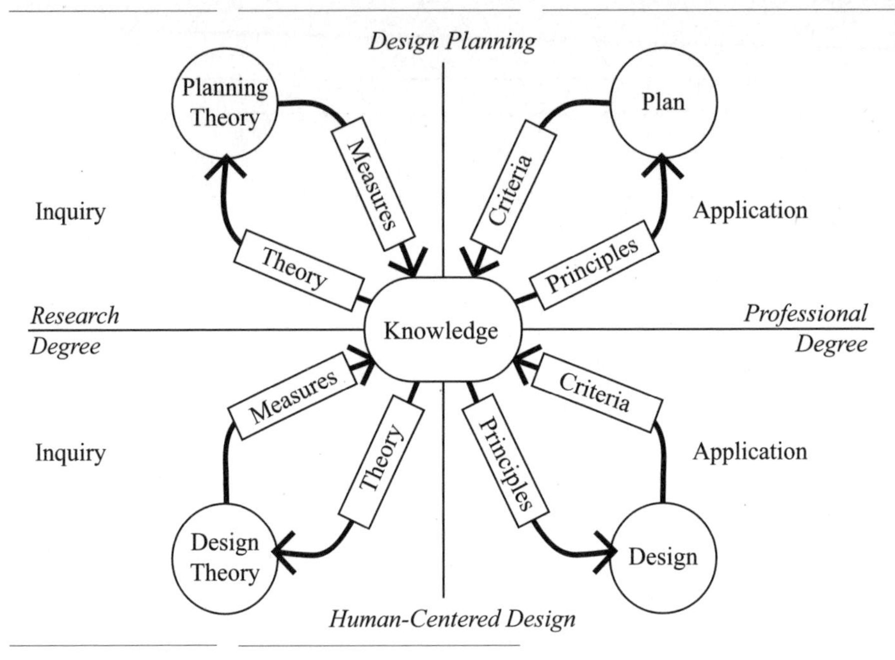

Figure 2
Graduate Study at the Institute of Design (Owen 1998).

> *Initiate studies of the philosophy of design. Just as studies of the philosophy of science, history, religion, etc. seek to understand the underpinning values, structures and processes within these systems of knowledge building and using, there need to be studies of the nature of design....*

which can be interpreted as the urge to do research ABOUT design.

Findeli combines two important aspects that have been mentioned before: firstly "the eclipse of the object," or the shift of research from artifacts (esthetics) to processes (logic) to experience (ethics), and, secondly, the essential role of project-grounded research for the development of a designerly research paradigm (Findeli 2008a). The concept shows a strong similarity to "Research THROUGH Design" (RTD).

Although it is reported that Archer first coined his phrase "... research about design [and designing], research through design [and designing] and research for the purposes of design [and designing]" in the late 1970s during his post at the Royal College of Art in London (http://www.core77.com/research/thesisresearch.html accessed 21 August 2008), it was Frayling (1993 → Text 5) who made the distinction popular. He introduced the categories of research *into / through / for* art and design.

We think that this categorization, which—for the first time—does not distinguish as to subject matter or an assumed structure of the "real world" but according to purpose,

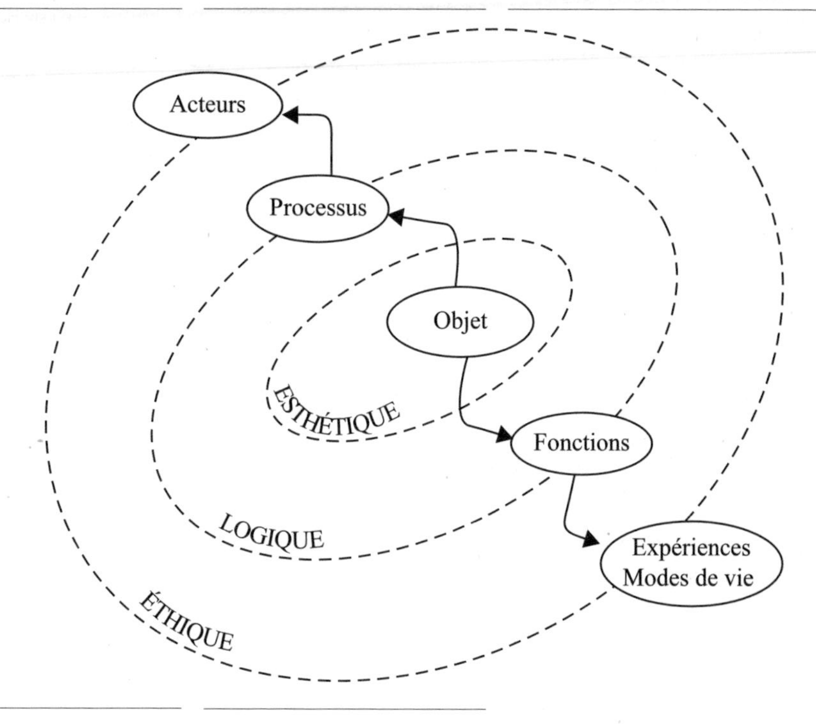

Figure 3
The "Bremen model" of design research spheres (Findeli 2005).

intentionality and attitude towards subject matters in design, is essential for a genuine designerly research paradigm. The inconsistencies seem to result from the obvious shift of meaning of *for* and *through* between Frayling (1993) and Findeli (1998), which need clarification and further elaboration (Jonas 2007a,b). See Table 5.

Research ABOUT and FOR design seems unambiguous. The epistemological status of Research THROUGH Design (RTD), however, is still weak. Grounded theory as well as action theory will probably contribute: *Grounded theory* is aiming at theory building, while accepting the modification of its subject matter. *Action research* is aiming at the modification of reality, while observing and processing theory modifications. Both approaches admit the involvement of the researcher as well as the emergence of theories from empirical data, in contrast to the established concept of theory building as the verification of previously formulated hypotheses.

Archer (1995 → Text 6) adheres to the distinction of research *about / for / through* design and puts RTD on a level with *Action Research* (1995, 11):

> *"It is when research activity is carried out through the medium of practitioner activity that the case becomes interesting."*

Table 5
Design research concepts based upon Frayling's (1993) terminology.

	into / about	for / for	through / by	remarks
Frayling 1993	into "... the most straightforward, and ... by far the most common: - Historical research - Aesthetic or Perceptual Research - Research into a variety of theoretical perspectives on art and design – social, economic, political, ethical, cultural, iconographic, technical, material, structural ... whatever. ... there are countless models – and archives – from which to derive its rules and procedures."	for "The thorny one is Research *for* art and design, ... Research where the end product is an artefact – where the thinking is, so to speak, *embodied in the artefact*, where the goal is not primarily communicable knowledge in the sense of verbal communication, but in the sense of visual or iconic or imagistic communication. ..."	through "... less straightforward, but still identifiable and visible. - materials research – such as the titanium sputtering or colorization of metal projects ... - development work – for example, customizing a piece of technology to do something no one had considered before, and communicating the results. ... - action research – where a research diary tells ... of a practical experiment in the studios, and the resulting report aims to contextualize it. Both the diary and the report are there to *communicate the results*, which is what separates *research* from the gathering of reference materials. ..."	Frayling's categorization is inconsistent and rather fuzzy: Frayling's "through" comprises much of Findeli's "for"; only action research may relate to "through / by" with both authors. Frayling's "for" is something very different from Findeli's "for"; Findeli would probably not consider it as research at all.
Findeli 1998	into / about Separation of design research and design practice (weak theory), "little or no contribution to a theory of design" see the field of "design studies" (Margolin)	for Design as applied science (no theory), complex, sophisticated projects (Research and Development)	by / through Conciliation of theory and practice (strong theory) embedded, implicated, engaged, situated (Sartre, Situationist) theory. "Such research helps build a genuine theory of design by adopting an epistemological posture more consonant with what is specific to design: the project."	Findeli's categorization provides an epistemologically and semantically much clearer concept.

Jonas 2004	about / über "Research *about* Design agiert von außen, den Gegenstand auf Distanz haltend. Forscher sind wissenschaftlich arbeitende Beobachter, die den Gegenstand möglichst nicht verändern. Beispiele: Designphilosophie, Designgeschichte, Designkritik, . . ."	for / für "Research *for* Design agiert ebenfalls von außen, den Prozess punktuell unterstützend. Forscher fungieren als "Wissenslieferanten" für Designer. Das bereitgestellte Wissen hat aber durchaus begrenzte Haltbarkeitsdauer, weil es sich auf eine durch Design zu verändernde Wirklichkeit bezieht. Beispiele: Marktforschung, Nutzerforschung, . . . Produktsemantik, . . ."	through/ durch "Research *through* Design bezeichnet das designeigene forschende und entwerfende Vorgehen. Designer / Forscher sind unmittelbar involviert, Verbindungen herstellend, den Forschungsgegenstand gestaltend. Beispiele: potentiell jedes 'wicked problem' im Rittelschen (1992) Sinne."	Jonas refers to Findeli's categorization.

Findeli (1998) argues that:

> *. . . "project-grounded research" . . . is a kind of hybrid between action research and grounded theory research, but at the same time it reaches beyond these methods, in the sense that our researchers in design are valued for both their academic and professional expertise, which is not the case even in the most engaged action research situations. . . .*
>
> *. . . although the importance of the design project needs to be recognized in project-grounded research, it should never become the central purpose of the research project, otherwise we fall back into R&D. Therefore, the design project and its output find their place in the annex of the dissertation, since practice is only a support for research (a means, not an end), the main product of which should remain design knowledge.*

The issue of rigor vs. relevance occurs again. Findeli (2008a) introduces a new perspective in arguing that research THROUGH design (or "project-grounded research" as he prefers to call it) has to combine research FOR and ABOUT design in order to become both relevant and rigorous. For the most current version, see Findeli (2008b → Text 7).

One might assert that we have a quite consistent, although still rough and not at all complete, conception of design research at this point. Scientific methods are necessary, but the nature of the design phenomena does not allow the reduction of design research to scientific research. On the contrary: scientific research has to be embedded in designerly models of inquiry. There are the all-embracing subject matters of *esthetics / products-logic / process-ethics / people*, and the essential distinguishing purposes of understanding design-relevant phenomena, of improving the design process, and of improving the human condition. These purposes have to be related to the epistemological attitudes of research ABOUT design—FOR design—THROUGH design.

3 Present-day pluralism and emerging patterns

Alongside the "designerly" strand, there are intensified attempts to re-align design research with scientific research, or to re-establish a clear distinction between (reflective) practice and "proper" research. For that purpose Friedman (2003 → Text 8) resorts to the established distinctions between clinical, applied, and basic research in medicine. He argues that while the dyadic division (basic-applied) may suffice for the natural sciences, it is not adequate for understanding research in the technical and social sciences or the professions they support.

The function of the distinction in the design context remains unclear. Medicine is about the improvement of the condition of the human body and mind, which means that there exists a more or less stable reference system for the measurement of success or failure. The distinction of clinical / applied / basic corresponds to the degree of decontextualization of the subject matter. Design always deals with the fit of highly fuzzy systemic wholes in real-world contexts, that immediately lose their significance in de-contextualized situations. Therefore one might argue that "basic" research is meaningless in design and that "clinical" research is the most "basic" and, at the same time, most challenging form of design research. Glanville (1980) and Archer (1995) support this view.

Friedman constructs an antagonistic distinction between reflection and research (2002):

> *. . . Design practitioners are always involved in some form of research, but practice itself is not research. While many designers and design scholars have heard the term "reflective practice," reflective practice*

Table 6
Friedman's distinction between research purposes in the framework of basic, applied, and clinical research (2003).

Basic research	. . . involves a search for general principles. These principles are abstracted and generalized to cover a variety of situations and cases. Basic research generates theory on several levels. This may involve macro level theories covering wide areas or fields, midlevel theories covering specific ranges of issues or micro level theories focused on narrow questions. Truly general principles often have broad application beyond their field of origin, and their generative nature sometimes gives them surprising predictive power.
Applied research	. . . adapts the findings of basic research to classes of problems. It may also involve developing and testing theories for these classes of problems. Applied research tends to be midlevel or micro level research. At the same time, applied research may develop or generate questions that become the subject of basic research.
Clinical research	. . . involves specific cases. Clinical research applies the findings of basic research and applied research to specific situations. It may also generate and test new questions, and it may test the findings of basic and applied research in a clinical situation. Clinical research may also develop or generate questions that become the subject of basic research or applied research.

is also not research, and reflective practice is not a research method as is sometimes mistakenly suggested. ...

What distinguishes research from reflection? Both involve thinking. Both seek to render the unknown explicit. Reflection, however, develops engaged knowledge from individual and group experience. It is a personal act or a community act, and it is an existential act. Reflection engages the felt, personal world of the individual. It is intimately linked to the process of personal learning (Friedman and Olaisen 1999; Kolb 1984). Reflection arises from and addresses the experience of the individual.

Research, in contrast, addresses the question itself, as distinct from the personal or communal. The issues and articulations of reflective practice may become the subject of research, for example. This includes forms of participant research or action research by the same people who engaged in the reflection that became the data. Research may also address questions beyond or outside the researcher. ...

What is significant about this, however, is that neither practice nor reflective practice is itself seen a research method. Instead, reflective practice is one of an array of conceptual tools used in understanding any practice—including the practice of research.

In short, research is the "methodical search for knowledge. Original research tackles new problems or checks previous findings. Rigorous research is the mark of science, technology, and the 'living' branches of the humanities" (Bunge 1999, 251). Exploration, investigation, and inquiry are synonyms for research. ...

We see the difference, but this desperate attempt at keeping up the barrier between the "swampy lowland" of reflective practice and the "high ground" of rigorous research is definitely a step backward when compared with the previous, however deficient, conceptual models of design research. Friedman's own words demonstrate the weakness of his postulate. In saying that *"reflective practice is one of an array of conceptual tools used in understanding any practice—including the practice of research,"* he implicitly states that reflective practice is an essential research medium, probably the most important one in the "Sciences of the Artificial." A circular one, admittedly.

Insisting on the clear distinction between "mere" design and "proper" research contributes to the "colonization of design" (Krippendorff 1994). "Mind the gap!" (Jonas and Meyer-Veden 2004) has been a polemic intervention against these scientistic positions. We should aim for the clarification of the systemic relation of Design-Research-Design Research.

Besides the "designerly" and the "scientific" strand, we see a plurality of schemes in peaceful co-existence. We mention just a few of them here:

Jones (no year → Text 9) takes the meta-perspective:

> *... the voice of reason asks ten questions:*
>
> *1. Is there a theory of design?*
> *2. What is the essential skill of designing—can it be described?*
> *3. What are design methods?*
> *4. How do you use them (design methods)?*
> *5. Is scientific research useful in designing?*
> *6. Is it scientifically possible to discover the nature of designing?*
> *7. How to design complex systems?*
> *8. How to solve or to avoid major problems created by the culture we create and inhabit?*
> *9. How to redesign the designed culture?*
> *10. How to teach this view of design—what are the principles?*
>
> *... to which the voice of intuition gives these spontaneous answers: ...*
>
> *5. Is scientific research useful in designing?*
>
> *When it comes to finding out how people experience and use existing or new designs, objective information is helpful, even essential—provided that the rigors of the laboratory and of scientific proof are not allowed to override well-informed intuitions but are used to replace ill-informed ones. Scientific methods, like all others, must be subject to an intuitive meta-method of navigation, such as I described in the previous answer. Without that they can be unproductive or destructive of insight and of life. There is no objective way to prove what is right.*
>
> *6. Is it scientifically possible to discover the nature of designing?*
>
> *There is a huge difficulty here. It seems that scientific research is suited only to observation, inductive theory, and experiment in relation to things that exist and that are separate from the people who are doing the research. Studying creative thought processes, such as design, is I think better done directly by introspection, by empathy, and by conversation, etc.*
>
> *The underlying difficulty of studying design is that it is concerned with the whole of something and "the whole" is not an objective reality—it*

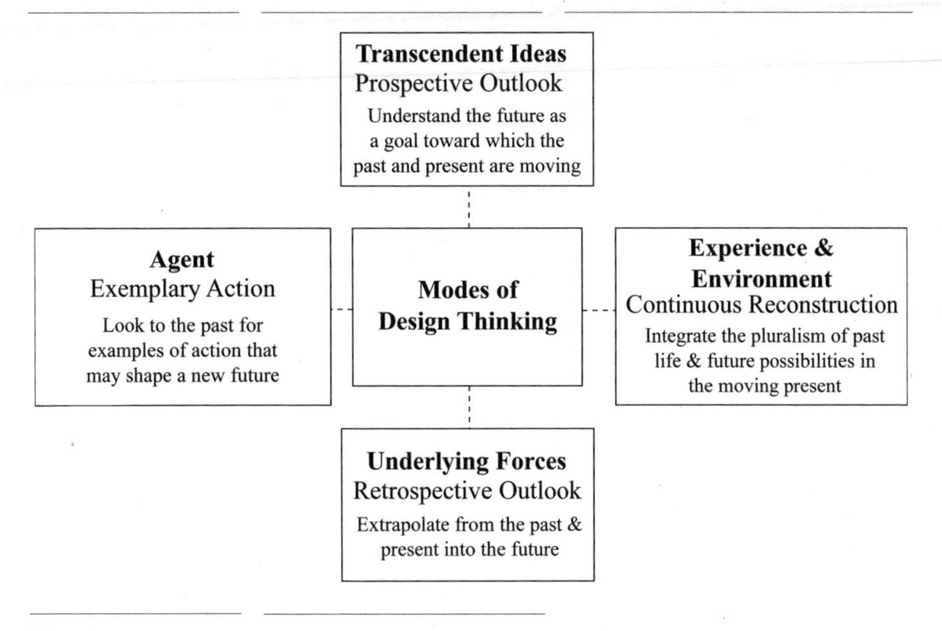

Figure 4
Four ways of theorizing about design (Buchanan 2001).

is a fluctuating scheme or state of the bodymind, perhaps more akin to religious inspiration than to science or to technology (William James, 1901–2). One can be totally involved. And what I am calling the bodymind includes not only the brain and body but one's conceptions of both body and of world. These are not external objects but they are surely realities.

Buchanan (2001) introduces a four-part scheme to explain the various ways of theorizing about designing and thus about researching design. According to Buchanan, there are four generative principles on which design theorizing stands: "Experience and environment," "Agent," "Underlying forces," and "Transcendent Ideas." The first two are oriented toward phenomenal processes and the last two toward ontic conditions.

Love (2002) introduces a hierarchy of subject matters relevant for design research that can be treated by means of established scientific tools.

Chow (2003) distinguishes five approaches according to underlying assumptions, regarding the nature of the process (1–3), regarding the intellectual approach (4), and regarding the purpose of designing (5). One might interpret this as research FOR (1–3), research ABOUT (4), and research THROUGH (5) design.

Sanders (2006) clusters the elements of the field in a four-quadrant scheme that focuses on the proportion of design vs. research in the process, and on the "mindset" of the designer

Table 7
Meta–theoretical hierarchy of concepts and theories in human activities (Love 2002).

Level	Classification
1	Ontological issues
2	Epistemological issues
3	General theories
4	Theories about human internal processes and collaboration
5	Theories about the structure of processes
6	Design and research methods
7	Theories about mechanisms of choice
8	Theories about the behavior of elements
9	Initial conception and labeling of reality

Table 8
Design research approaches according to Chow (2003).

	approach	representatives
1	Cognitive Problem Solving	Simon, Schön, Cross, . . .
2	Knowledge Processing	Bertola, Cooper, Reinmoeller, Tellefsen, . . .
3	Communication Interaction	Broadbent, Jonas, Findeli, Pizzocaro, . . .
4	Philosophical Intellectual	Buchanan, Sless, Nelson & Stolterman, . . .
5	Radical Democratic	Fry, Jones, Manzini, Margolin, . . .

or researcher, defined as expert vs. participatory. The essential aspect is that there are no clear distinctions between design and research anymore.

Langrish (14 Dec. 2007 on DRS–list) emphasizes the evolutionary character of design and design research and indicates four strands that are somewhat U.K.–specific, and offers the comforting advice that we should not worry about the diversity / messiness of concepts and approaches:

> *Design Research, like most human activities, can only be understood in evolutionary terms. That means it has a history but its future is not predictable and there are some very silly things that survive. Biology has the duck-billed platypus and academia has the PhD, which is VERY silly. There are four strands to the history of design research.*
>
> *1. Mechanical Engineers with an interest in engineering design and the desire to make the process more rational.*

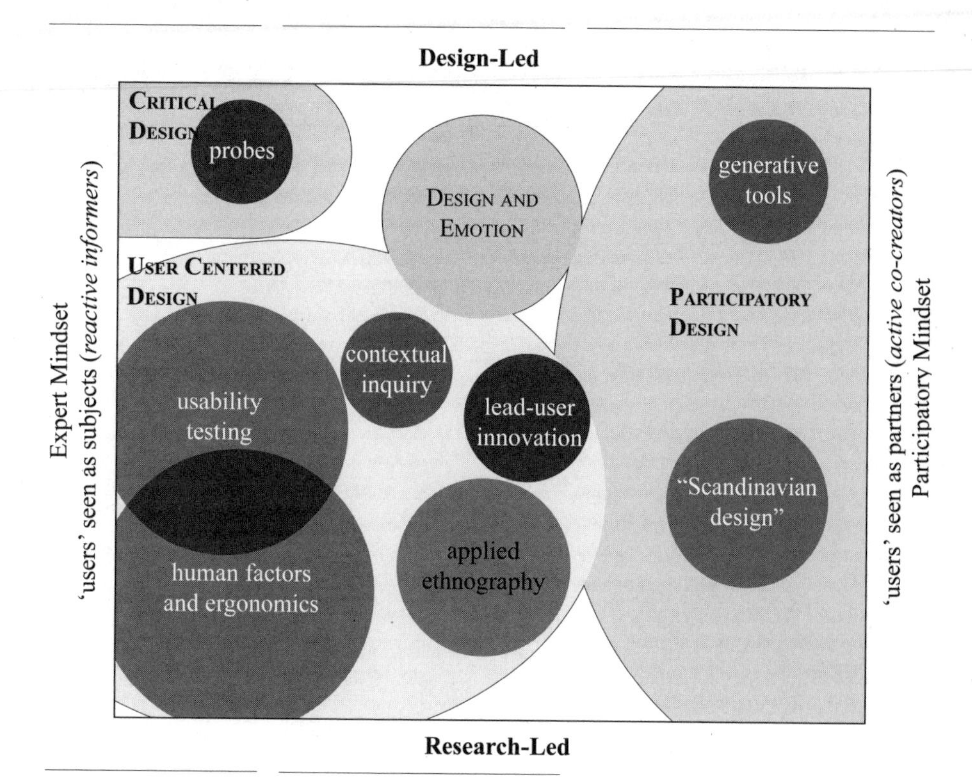

Figure 5
Topography of Design Research (Sanders 2006).

2. Architects who have always had people who speculated about what does it all mean and how do we do it. This goes back to Vitruvius and his books on architecture, which, as well as having practical stuff, also speculate about the origins of different styles, etc.

3. Historians and later cultural studies-type people who are outsiders looking in—along with a few psychologists and the odd sociologist.

4. Late on the scene were the old art colleges, absorbed into "Polys" and then becoming universities and finding themselves with the research assessment exercise. Within present people's lifetimes, this sector had to work out what is an honors degree, then what is a Master's, and then cope with PhDs against the background of an educational activity which encouraged creativity and discouraged scholarship. So you still find people arguing about what is research and getting VERY confused about research and practice.

There is nothing unusual about a ragbag of different kinds of questions with different methods for trying to answer them. Chemistry, for example, is four quite different subjects which historically came together and each of them has many subdivisions. (It's only a historical accident that stopped electricity from being part of chemistry.) Physical chemists look down on organic chemists as "cooks." Organic bods think the physical bods have no feel for chemistry—they are just adding machines, and so on.

So don't worry about it—just enjoy doing whatever you do.

John Z L

Dorst (2008) emphasizes the need for more integrative views, and argues that *"a revolution [is] waiting to happen"* in design research. He states that research has frequently lost contact with practice in recent years, and that there has been too much focus on process, neglecting the designer, the task, and the context in design research. Therefore, he pleads for re-engagement with practitioners and for "complexification" of design research. In view of the above considerations, we cannot share his position and argue instead that the "revolution"—probably rather an evolution—has been going on for several decades and that viable patterns are emerging. For instance, subject matters and purposes seem uncontroversial. However, the epistemological issues, which are closely related to methodology, still require clarification.

4 Methodology and epistemology: three phases and four categories

If we take a more general and slightly fuzzy designerly view, a striking pattern emerges from this messy picture. Hybrid and integrative design and research process models are evolving that acknowledge the "beauty of gray" between "mere" design and "proper" research, and argue in favor of a specific epistemological status for design research. A genuine design-specific triadic structure of the process, albeit in different terminologies, is emerging in various "sciences of the artificial" (or "making disciplines," or disciplines dealing with the teleological transfer of an existing state into a preferred one), such as *design* (Jones 1970; Archer 1981; Nelson and Stolterman 2003; Jonas 2007), *management* (Weick 1969; Simon 1969), and *HCI* (Fallman 2008).

The interpretation is still contentious. Is it a necessary sequence or a loose combination of different spheres of knowing? Fallman (2008) argues that the same tools are used in the three fields, but for different applications. Jonas argues that Research THROUGH Design (RTD) means the integration of all three components. In any case, the middle column seems to be the central design-specific component, where the essential "designerly" competencies are located. Design epistemology develops towards methodological patterns that show a resemblance to basic structures in transdisciplinarity (Nicolescu 2002) studies, which strive for an integration of system knowledge, target knowledge, and transformation knowledge. Every

Table 9
Triadic concepts / domains of knowing in design research, indicating a generic tripartite model of the designerly research process (see also Chow 2009).

author	phases / components / domains of design research		
Jones 1970	divergence	transformation	convergence
Archer 1981	science	design	arts
Simon / Weick 1969	intelligence	design	choice
Nelson & Stolterman 2003	the true	the ideal	the real
Jonas 2007	analysis	projection	synthesis
Fallman 2008	design studies	design exploration	design practice

design process follows this generic cybernetic structure (see below), making use of the various scientific and designerly methods provided for each of the steps. The inherent fuzziness of the process model makes it possible to bridge the causality gaps occurring between the different, often incompatible, scientific contributions.

The emerging paradigm of RTD argues that it is the generic design process and not the scientific process that guides design research. The archetypal anthropological pattern of acting combined with retrospective as well as projective reflection, which—later in history—is called "designing," is the essential characteristic of being human. This "naturalistic" approach does not resort to any ontological or metaphysical preconditions except the belief that designing is an evolutionary achievement. The scientific paradigm must be embedded in the design paradigm:

- *design research is guided through design process logic,*
- *design research is supported by phases of scientific research and inquiry, and*
- *complexity must not be unduly reduced, or the subject matter of design research is destroyed.*

These considerations support the hypothesis of two central constituents of the theoretical framework for mapping design research: *cybernetic learning models* (accounting for evolutionary properties) and *second order cybernetic models of observer / actor / designer involvement* (accounting for systemic complexity).

4.1 Three phases: First-order cybernetic learning models integrate design and science

In epistemological terms, designing can be seen as a cybernetic *learning process*, one which is biologically grounded in the need of living systems to survive in an environment. Learning is conceived as a feedback cycle of acting and reflecting. The aim is not final "true" representation of some external reality, but rather a process of *(re-) construction* for the purpose of appropriate *(re-) action*. Evolutionary epistemologists (Campbell 1974) argue that the Kantian transcendental *a priori* must be replaced by the assumption of an *evolutionary fit* between the objects and the subject of recognition. The well-known circular design process models, such as that of the Institute of Design Chicago (research—analysis—synthesis—realization) seem to be adaptations of Kolb's (1984) "learning cycles." The latter, in turn, seems to be an adaptation of the very basic cybernetic O.O.D.A. model of the U.S. Air Force.

Innovation is about novelty generation, or the creation of new unexpected stable objects or forms, of in-form-ation (Glanville 2008). This has often been neglected in design research. Where is the link between Simon's state 1 and state 2, between inductive hypothesizing and deductive concluding? The logical syllogisms are clearly unable to explain the generation of new facts and artifacts. How is the scientific hypothesis generated? How is the design concept created? *Induction* and *deduction* are accepted syllogisms in science and design, but they do not explain the creation of the new. Based upon pragmatist concepts from Peirce (Davis 1972), Dewey (1986), and others, we consider *abduction* to be the central mental and social "mechanism" of knowledge generation in general, applicable in everyday life, in the designerly as well as in the scientific process. It is the abduction step that is able to combine the otherwise sterile syllogisms of induction (formulating a rule out of existing data or cases) and deduction (deriving special cases from rules) into a productive learning cycle. Without abductive reasoning, "normal science" (Kuhn 1973) is the most that would be possible. March (1984) states:

> *As Peirce writes: abduction, or as we have it production, "is the only logical operation which introduces any new ideas; for induction does nothing but determine a value; and deduction merely evolves the necessary consequences of a pure hypothesis." Thus, production creates, deduction predicts; induction evaluates.*

Roozenburg (1993) renders these considerations more precisely. He differentiates between *explanatory abduction* and *innovative abduction*, and concludes that it is the latter that should be taken as the "paradigm" model of the crucial step in the design process that generates the new:

In explanatory abduction it is assumed that the rule (of the syllogism) is given as a premise; innovative abduction aims at finding new rules....

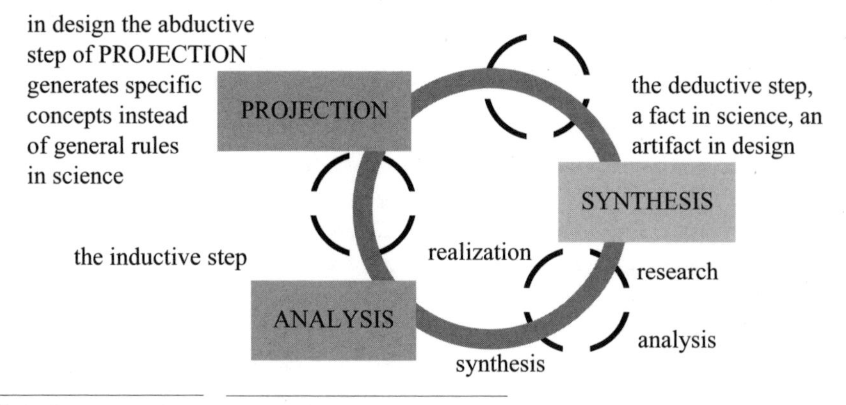

Figure 6
A cybernetic learning model of 3 macro phases and 4 micro steps (Chow and Jonas 2008).

In designerly methodological terms we speak of ANALYSIS (the inductive phase), PROJECTION (the abductive phase), and SYNTHESIS (the deductive phase). PROJECTION is essential for bridging the gap: in science the gap is finally removed by means of a generalized logical construction, and PROJECTION can remain a mystery. In design, the gap is temporarily bridged; the art of PROJECTION is the essential task. Thus the further clarification of the abductive mechanisms in the PROJECTION phase is crucial for the development of genuine designerly concepts of research and the subsequent spreading of those concepts to other fields. If we combine two circular learning models, the macro model of ANALYSIS—PROJECTION—SYNTHESIS (the domains of knowing, see Table 9) and the micro model of research → analysis → synthesis → realization (the learning steps), we obtain a hypercyclic *generic design process model* (Hugentobler, Jonas, Rahe 2004).

This establishes the model of RTD , which seems closely related to mode 2 science. We argue that it is the PROJECTION phase that integrates science and design.

Table 10
RTD performs an integrative function by means of PROJECTION, which links design and scientific research.

	ANALYSIS the true	PROJECTION the ideal	SYNTHESIS the real
Design			
RTD (mode 2 science)			
Scientific Research (mode 1 science)			

4.2 Four categories: second-order cybernetic models of observer involvement account for complexity

As mentioned above, distinctions concerning disciplinary subject matters do not make sense in design research, because design always deals with systemic wholes consisting of hybrid conglomerates of diverse elements. We need distinctions that deal with the positions, intentions, or attitudes of the observers / designers instead. Any purposive process such as design has to reflect the observer's involvement, meaning his / her positions in relation to the designing / inquiring system. The move to *operational epistemology* (von Foerster 1981), or from first- to second-order cybernetics has to be made eventually.

First-generation design methodology, as generally conceived, provides normative methods FOR the design process. This is a seemingly scientific attitude, one which, however, neglects the researcher's involvement and the dynamic context of every design research task. One indisputable merit of first-generation methodology research in the 1960s was that generic process models were considered in some depth. The notorious criticism of their rigidity is justifiable only when these models are considered as normative standards. The second-order cybernetic approach of reflecting observation modes (Glanville 1997) brings more clarity. It provides an explanatory basis for the concept of research FOR / ABOUT / THROUGH design, and reveals a new "inaccessible" category (Chow and Jonas 2008).

Observer position / ● looking →	outside the design system first-order cybernetics	inside the design system second-order cybernetics
outwards	research FOR design research based upon certain assumptions regarding the structure / nature of design processes, aiming at their improvement	research THROUGH design research guided by the design process, aiming at transferable knowledge and innovation
inwards	research ABOUT design research by means of disciplinary scientific methods, applied in order to explore various aspects of design	INACCESSIBLE (research AS design?) probably the essential mental and social "mechanism" of generating new ideas, the location of abductive reasoning

Figure 7
Knowledge generation in design research: the concepts of research FOR / THROUGH / ABOUT design in relation to observer positions (Glanville 1997).

Glanville (1997) explains this as follows:

ABOUT: The observer is outside, looking inwards:
The observer can attribute stability to the system (or the goal, see below). In doing so, he must attribute an internal goal. In defining the goal he becomes part of a larger (super)system. The attribution of stability to a system does not mean it behaves with pattern: pattern is an intentional result of fortunate observation. Its behavior is best described as random.

FOR: The observer is outside, looking outwards:
The observer can determine the goal and attribute purpose (relationship) between the system and the goal. But in doing so he becomes part of a larger system: that is, a new system is created, with the observer inside it (all is made within). Both system and goal are considered stable, and there is a firm (purposive) behavior of the system vis-a-vis the goal.

THROUGH: The observer is inside, looking outwards:
The observer is a steersman. He can choose a goal and must also separate himself from the system he is taken to be embedded in by the observer who is outside the system looking inwards. In assuming the goal and the system are stable, he must attribute to them goals. He gives purpose to the behavior (movement) of the system towards the goal: in this, in having separated himself, he is fulfilling the same role as the observer outside, looking inwards. Thus observation transcends boundaries.

AS?: The observer is inside, looking inwards:
We cannot, as outsiders, speak of this. We can deduce the need for self-reference, and forms such as autopoiesis (Maturana, Varela, and Uribe), the Paskian concept (Pask 1975) and my own Object (also von Foerster's token objects (von Foerster 1981), which have the interesting quality of appearing to recursively attain towards a state of stability, apparently through repetitive internal dynamics—although the reference point by which this stability can be estimated is nevertheless external). It is more dramatic, and more in keeping with the intention of this paper, to leave this blank, here.

Findeli (2006) identifies comparable modes, albeit using a different terminology:

- *AS: "Premier type : modèle de la théorie minimale,"*
- *ABOUT: "Deuxième type : la théorie comme cadre interprétatif,"*
- *FOR: "Troisième type : le design comme science appliquée,"*
- *THROUGH: "Quatrième type : le design comme théorie située et pratique éclairée."*

We conclude that research in design only makes sense when all observation modes are taken into consideration. Otherwise, the process remains locked in sterile assumptions that prevent the productive use and further dynamic development of methodology THROUGH design. It is the INACCESSIBLE (?) abduction step that is able to combine the logical syllogisms of induction (formulating a rule out of existing data—post-rationalization) and deduction (deriving special cases from rules—pre-rationalization) into a productive cycle with the potential for creating the new. Here we see the "fractal character" of the endeavor: of course, RTD requires "objective" scientific input generated by research FOR or ABOUT design. This playful dance of perspectives is what we consider second-generation methodology, which is, in our view, the most important conversational medium for the generating of new design knowledge. It takes into account the problem of irreducible complexity in design research situations (Mikulecky):

> *Complexity is the property of a real world system that is manifest in the inability of any one formalism being adequate to capture all its properties. It requires that we find distinctly different ways of interacting with systems. Distinctly different in the sense that when we make successful models, the formal systems needed to describe each distinct aspect are NOT derivable from each other.*

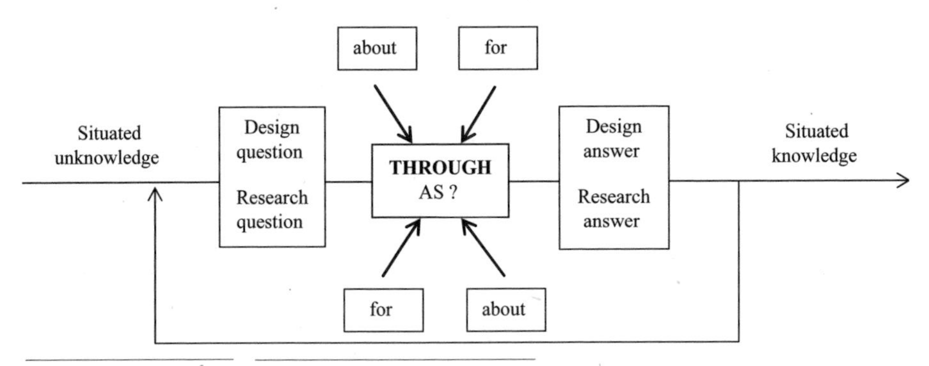

Figure 8
Research THROUGH design means the reflected, purposive, and playful use of observer positions during the design research process.

The final scheme of four categories of design research follows these considerations:

Table 11
Observer perspectives in the design research process (see Fig. 7).

Research is defined / determined by underlying basic assumptions regarding the design process (What is design? How does it work?) aiming at its improvement. → Design as: - cognitive process - semiotic process - communicative process - learning process - . . . Emphasis on the analytic aspects of the research / learning cycle → research FOR design	Research is defined / determined by basic assumptions regarding the purpose of designing (What is design good for?) aiming at the achievement of goals. → Design as: - projective process - human-centered process - innovation process - political / social process - . . . Emphasis on the synthetic aspects of the research / learning cycle → research THROUGH design
Research is defined / determined by motivations aiming at inquiring into and understanding the "nature" of diverse aspects of design → Design as subject of disciplinary research: - historical - philosophical - psychological - . . . → research ABOUT design	Research in action . . . → Design as the INACCESSIBLE medium of knowledge production . . . → research AS design

Epilogue

Figure 9
Alice entering the room behind the looking–glass . . . the INACCESSIBLE / research AS design.

In another moment Alice was through the glass, and had jumped lightly down into the Looking–glass room. The very first thing she did was to look whether there was a fire in the fireplace, and she was quite pleased to find that there was a real one, blazing away as brightly as the one she had left behind. "So I shall be as warm here as I was in the old room," thought Alice: "warmer, in fact, because there'll be no one here to scold me away from the fire. Oh, what fun it'll be, when they see me through the glass in here, and can't get at me!"

Then she began looking about, and noticed that what could be seen from the old room was quite common and uninteresting, but that all the rest was as different as possible. For instance, the pictures on the wall next the fire seemed to be all alive, and the very clock on the chimney-piece (you know you can only see the back of it in the Looking–glass) had got the face of a little old man and grinned at her.

"They don't keep this room so tidy as the other," Alice thought to herself, as she noticed several of the chessmen down in the hearth among the cinders; but in another moment, with a little "Oh" of surprise, she was down on her hands and knees watching them. The chessmen were walking about, two and two!

. . .

(Carroll 1996, 135–136)

REFERENCES

Archer, Bruce (1979) "Design as a Discipline" in *Design Studies* 1 (1): 17–20.

Archer, Bruce (1981) "A View of the Nature of Design Research" in Jacques, R. and Powell, J. (eds.), *Design:Science:Method*, Guildford: Westbury House.

Archer, Bruce (1995) "The Nature of Research" in *Co-design* January: 6–13.

Bayazit, Nigan (2004) "Investigating Design. A Review of Forty Years of Design Research" in *Design Issues*, 20 (1), Winter: 16–29.

Buchanan, Richard (2001) "Children of the Moving Present. The Ecology of Culture and the Search for Causes in Design" in *Design Issues* 17 (1): 67–84.

Campbell, D.T. (1974) "Evolutionary epistemology" in Schlipp, P.A. (ed.) *The Philosophy of Karl Popper* Vol. 1: 413–463, La Salle, IL: Open Court Publishing.

Carroll, Lewis (1996) *The Complete Illustrated Lewis Carroll*, Hertfordshire: Wordsworth Editions.

Chow, Rosan (2003) "Shop around: design theories for design education," ICSID 2nd Educational Conference. Critical Motivations and New Dimensions, Hannover: iF International Forum Design GmbH. http://www.designresearchnetwork.org/drn/files/Chow_Shop%20around.pdf

Chow, Rosan and Jonas, Wolfgang (2008) "Beyond Dualisms in Methodology—an integrative design research medium ('MAPS') and some reflections," DRS conference *Undisciplined!*, Sheffield, 07.

Chow, Rosan (2009) "Fallman meets Jonas," *Communication by Design*, Sint-Lucas, 15–17 April, Brussels.

Cross, Nigel (1999) "Design research: A disciplined conversation" in *Design Issues* 15 (2), Spring: 5–10.

Cross, Nigel (2001) "Designerly Ways of Knowing: Design Discipline Versus Design Science" in *Design Issues* 17 (3), Summer: 49–55.

Davis, W. H. (1972) *Peirce's Epistemology*, The Hague: Martinus Nijhoff.

Dewey, John (1986) *Logic: The Theory of Inquiry*, Carbondale IL: Southern Illinois University Press.

Dorst, Kees (2008) "Design Research: A Revolution Waiting To Happen," in *Design Studies* 29 (1).

Fallman, Daniel (2008) "The Interaction Design Research Triangle of Design Practice, Design Studies, and Design Exploration" in *Design Issues* 24 (3): 4–18.

Findeli, Alain (1998) "A Quest for Credibility: Doctoral Education and Research in Design at the University of Montreal" in *Doctoral Education in Design*, Ohio, Oct. 8–11.

Findeli, Alain and Bousbaki, Rabah (2005) "L'éclipse de l'objet dans les theories du projet en design," *The Design Journal* VIII (3): 35–49.

Findeli, Alain (2006) "Qu'appelle-t-on 'théorie' en design? Réflections sur l'enseignement et la recherche en design" in Brigitte Flamand, *Le design. Essais sur des théories et des pratiques*, Paris: Edition du Regard.

Findeli, Alain (2008a) "Research through Design and Transdisciplinarity: A Tentative Contribution to the Methodology of Design Research" in *Proceedings of Focused, Swiss Design Network Symposium 2008*, Berne, Switzerland, 67–91.

Findeli, Alain (2008b) "Searching for Design Research Questions," keynote at *Questions & Hypotheses*, Berlin, October 24–26, 2008.

Frayling, Christopher (1993) "Research in Art and Design" in *Royal College of Art Research Papers* 1 (1): 1–5.

Friedman, Ken (2002) "Theory Construction in Design Research. Criteria, Approaches, and Methods" in *Common Ground. Proceedings of the Design Research Society International Conference at Brunel University, September 5-7, 2002.* David Durling and John Shackleton, Eds. Stoke on Trent, UK: Staffordshire University Press.

Friedman, Ken (2003) "Theory construction in design research: criteria, approaches, and methods" in: *Design Studies* 24: 507–522.

Fuller, Richard Buckminster (1999) *Utopia or Oblivion,* New York: Bantam Books.
Glanville, Ranulph (1980) "Why Design Research?" in: Jacques, R. and Powell, A., *Design: Science: Method,* Guildford: Westbury House.
Glanville, Ranulph (1997) "A Ship without a Rudder" in: Glanville, Ranulph and de Zeeuw, Gerard, eds. *Problems of Excavating Cybernetics and Systems,* BKS+, Southsea.
Jacques, R. and Powell, A. (eds.) (1981) *Design: Science: Method, Proceedings of the 1980 Design Research Society Conference,* Guildford: Westbury House.
Jonas, Wolfgang (1999) "On the Foundations of a 'Science of the Artificial'" in *International Conference on Art and Design Research,* University of Art and Design, Helsinki UIAH, Finland, September 9–11.
Jonas, Wolfgang (2000) "The paradox endeavour to design a foundation for a groundless field" in *Re-Inventing Design Education in the University,* International Conference December 11–13, Perth, Australia.
Jonas, Wolfgang (2003) "Mind the gap! – on knowing and not-knowing in design. Or: there is nothing more theoretical than a good practice" in *Proceedings of Design Wisdom—techné, the European Academy of Design*, Barcelona, Spain, April, 28–30.
Jonas, Wolfgang and Meyer-Veden, Jan (2004) *Mind the gap! On knowing and not-knowing in design*, Bremen: Hauschildt.
Jonas, Wolfgang, Chow, Rosan, and Verhaag, Niels (2005) *Design – System – Evolution, Proceedings of the 6th conference of the European Academy of Design, University of the Arts,* Bremen, March 29–31. http://ead.verhaag.net/conference/
Jonas, Wolfgang (2007a) "Design Research and its Meaning to the Methodological Development of the Discipline" in *Design Research Now—Essays and Selected Projects,* Basel: Birkhäuser Verlag.
Jonas, Wolfgang (2007b) "Research through DESIGN through research—a cybernetic model of designing design foundations" in *Kybernetes* 36 (9), special issue on cybernetics and design.
Jones, John Christopher (no year), "a theory of designing," see http://www.softopia.demon.co.uk/2.2/theory_of_designing.html, accessed Feb. 25, 2010.
Kolb, David A. (1984) *Experiential learning: experience as the source of learning and development*, New York: Prentice-Hall.
Krippendorff, Klaus (1994) "Redesigning Design. An Invitation to a Responsible Future" in *Design—Pleasure or Responsibility? International Conference on Design at the University of Art and Design*, Helsinki UIAH, June 21–23.
Krippendorff, Klaus (2007) "Design Research, an Oxymoron?" in *Design Research Now—Essays and Selected Projects*, Basel: Birkhäuserr Verlag.
Kuhn, Thomas (1973) "Die Struktur wissenschaftlicher Revolutionen," Frankfurt/Main: Suhrkamp.
Latour, Bruno (1998) *Wir sind nie modern gewesen. Versuch einer symmetrischen Anthropologie*, Frankfurt / Main: Suhrkamp.
Love, Terence (2002) "Constructing a coherent cross-disciplinary body of theory about designing and designs: some philosophical issues" in *Design Studies 23:* 345–361.
March, L. (1984) "The Logic of Design" in Cross, Nigel (ed.), *Developments in Design Methdology,* Chichester: John Wiley, 265–276.
Mikulecky, D.C. (no year) "Definition of Complexity," see http://www.people.vcu.edu/~mikuleck/ON%20COMPLEXITY.html, accessed April 10, 2010.
Nelson, Harold G. and Stolterman, Erik (2003) *The Design Way. Intentional Change in an Unpredictable World*, Englewood Cliffs, NJ: Educational Technology Publications.
Nicolescu, Basarab (2002) *Manifesto of Transdisciplinarity,* Albany NY: State University of New York Press.

Nowotny, Helga, Scott, Peter, and Gibbons, Michael (2001) *Re-Thinking Science. Knowledge and the Public in the Age of Uncertainty,* Cambridge, UK: Polity Press.
OODA, see http://en.wikipedia.org/wiki/OODA_loop, accessed Feb. 25, 2010.
Owen, Charles (1998) "Design research: building the knowledge base" in *Design studies* 19: 9–20.
Rittel, Horst W. J. (1972) "Second-generation Design Methods" in Cross, Nigel (ed.) (1984) *Developments in Design Methodology,* Chichester: John Wiley, 317–327. (1st ed. 1972.)
Roozenburg, N. F. M. (1993) "On the Pattern of Reasoning in Innovative Design" in *Design Studies* 14 (1): 4–18.
Sanders, Liz (2006) "Design research in 2006" in *DesignResearchQuarterly* V.I:1 September: 1–8. http://www.designresearchnetwork.org/drn/files/Sanders-Cluster.jpg
Schön, Donald A. (1983) *The Reflective Practitioner. How Professionals Think in Action,* New York: Basic Books.
Simon, Herbert A. (1969, 1981, 1996) *The Sciences of the Artificial,* Cambridge MA: MIT Press.
Undisciplined! (2008) Conference of the Design Research Society, Sheffield, 2008.
Weaver, Warren (1948) "Science and Complexity" in *American Scientist* 36: 536–544.
Weick, Karl (1969) *Social Psychology of organizing,* Reading MA: Addison Wesley.

Ranulph Glanville, Independent Academic[1]

1 RE-SEARCHING DESIGN AND DESIGNING RESEARCH[2]

Prologue[3]

When Design Research began, say in the 1960s, the eventual success of science was assumed. Already, at the notorious 1956 Oxford Conference, architectural education in the UK (and its sphere of influence) accepted architecture was a second class subject: i.e, not properly scientific. Science (in actuality, technology) was seen as so successful that everything should be scientific: the philosopher's stone! Architects (a significant subdivision of designers) were determined to become scientific. The syllabus was changed and design science was invented. Even the Architectural Association School gave over a third of undergraduate time to design science. Prime Minister Wilson and his Government declared the "White Heat of the Technological Revolution."

It was no wonder design was seen not as a discipline in its own right. Design was deficient: effectively, a defective science. It was flawed. But these flaws could be fixed by the proper application of scientific methods.

[...]

Research was what was needed. Proper scientific research (research was identified with science) would yield the secrets of the designer, allowing us unsentimentally to find the right answer to problems. Research was central to Science. Research was Science. In shameful contrast, Design was not Scientific. Design should be Scientific. Design therefore needed Research.[4] Since Research should be Scientific, Design Research should be Scientific. And then Design, itself, would be Scientific.

Research gave Science its precedence, even today indicated in the growth in evidence-based studies.

In this paper, I look at research, both experimental and theoretical, as done compared with as reported, and compare this to what is central to the act of design (as I understand it) to throw light on the relationship that could hold between research and design. Thus, the balance may be restored so design is accorded what I consider its proper position. Finally, I consider whether there is an area of knowledge which has already achieved, within its own competence, what design should aim to achieve for itself.[5]

So the purpose of this paper is to construct an argument that gives design back its rightful place in research: that is, shows research to be a (restricted) design act, rather than design being an inadequate research.[6]

Part I: the World of Research

Research

What is the purpose of research? What do we aim to achieve through it?

Research is an undertaking by which we aim to increase our knowledge (of the world).[7] The word (and the word design) is both noun and verb in English. In this paper, I am specially interested in research and design as activities, so generally my use is of the verb.

Research is usually understood to produce extendable and testable social knowledge. A characteristic (Swanson 1997) is that we take our knowledge, extend and test it until it "breaks" and then rebuild it. Thus we extend what we know. The circularity and failure (leading to "rebirth") is central to the research undertaking.

What research produces—the outcome—should be stable to be useful in making knowledge, i.e., the outcome should be repeatable unambiguous (stable in interpretation).

It should also be coherent: the outcome should fit with (occasionally cause reconsideration of) what is already known. Research is concerned with both individual chunks of knowledge, and their assembly into larger structures. It is important that the chunks stick together within the larger structures. This implies that coherence is deeply connected with consistency: the chunks must be consistent with eachother within the structures.[8]

An important way of determining that our knowledge is consistent and repeatable (i.e, complete) is in predicting outcomes. When knowledge does this successfully, we extend our belief in it, leading to a science of conjectures. The fact that we are willing (in theory, at least) to test these conjectures to destruction leads to a science of refutations in Popper's manner (1969).

However, scientific research is not always carried out according to Popper's ideal, which is impossibly ambitious for mere humans. Kuhn (1970) argued from historical observations. He divided science into the revolutionary, which is distinct from the more mundane and technical tasks of conserving the status quo that he calls normal science. Lakatos (1969) has indicated the development around accepted theories of "protective belts" that repel the unconventional.

Science as practised is not the ideal Popper suggests. We may aim for, but are not likely to attain, Popper's ideals. This difference between science as portrayed and done is important.

Research is carried out in two main arenas. The first is experiment; the second, theory. I am concerned with the actuality of what happens, in contrast to the "official presentation."

All this happens against a background of assumptions, for instance that something has always happened does not mean it always will. But we assume that the likelihood of it continuing to happen increases the more it has already happened, until. . . .[9]

Experiment

Experiments are the main means by which scientists extract knowledge of the world we inhabit.[10] They do this by radical simplification.

In the (idealised) scientific experiment, we divide systems into distinct, isolated variables. We fix all but one of the variables, and change some factor we believe is influential in the behaviour of the system, realised in the free variable. Changing this factor, we observe any change in the behaviour of the system and attribute it to the response of that variable. We organise the "in and outputs" so that there appears to be a simple relationship, and we determine that this relationship is determined by the variable.[11]

We have devised methods (e.g., statistics) for "faking" these conditions in complex systems where we cannot isolate variables, and/or where repeatibility is unattainable.

I am sure the reader is familiar with the above picture. What is left out is the experimenter. Yet how could there be an experiment without an experimenter?

The answer (based on simple experience: of doing experiments, and of language use and what that implies for action) is straightforward. We cannot.

The experimenter chooses to do the experiment and sets it up (including determining the variables).[12] The experimenter observes, and determines what the outcomes are. The experimenter carries out the actions.

The experimenter continues until the system begins to perform as desired or required (for instance, moves the light source/screen/lens to get an in-focus image). And the experimenter determines when enough has been done, i.e., who breaks the circle.

The experimenter designs the experiment, and if it doesn't work (well enough, in his opinion) redesigns it. The experimenter forms the outcome and assembles different observations into a coherent whole (relating them together). The experimenter may then tie the outcome into theory, modifying that accordingly. Reacting to changes in knowledge resulting from the experiments, the experimenter may rerun the experiment, perhaps with a new arrangement of the variables, or in a different place, to check repeatibility (i.e., stability). He may make predictions (which always requires a rerun). The experimenter plays with all aspects of the experiment until it produces results of the type desired.

These actions of the experimenter are circular. As a result of their circularity, novelty (the unexpected) may be observed, leading to a rerun under changed circumstances.

There are circularities in setting up and doing the experiment, in valuing what is found and integrating it. There are circularities of repetition. The whole process is deeply embedded in circularity, particularly the greatest of all scientific circularities: the active involvement of the experimenter (the observer).

Account

[...] The undertaking is presented in a manner supporting the view that The Great Scientific Endeavour brings forth truths unsullied by human intervention, awaiting discovery in The Great Reality Out There. Nowadays, this position is unreasonable and untenable: it is hard to pursue the consequences to a logical conclusion, and take onboard the constructivist view that seems to emerge. In science, we have long been drilled otherwise.

[...]

Over the last thirty years, linguistic analysis indicates the aims of the scientific communication and the use of language indicating this have changed. Publications are no longer concerned with "the truth"; they communicate the author's (or the editor's) wish to join or remain in a group of fellow workers (Hunston 1993; Hyland 1997; Glanville, Forey, and Sengupta, forthcoming). This finding holds for physicists as for social scientists. The first person has found expression again: papers are written by I's. Designers know they are not dealing with "the truth" except metaphysically, e.g., truth to materials (the Architectural Association's motto is "Design in Beauty, Build in Truth"): and they know there is no design without them, the designers.

We also nowadays understand that the description is not the experience; the explanation is not the actuality; prediction is not mechanism.[13]

Finally, we fail to mention the actual processes of writing/reporting—as above!

Theory

Theory is what turns the (collections of) observations we build into science. It may not make these collections science, but, if not sufficient, it is necessary. However, this paper is not primarily concerned with what makes science, but the role of theory in research.

In my understanding, theory in research has the following two roles.

> First, to combine, co-ordinate and simplify the findings of experiments by developing generalising concepts.
>
> Second, to examine these concepts in order to further clarify and develop them, reflecting back extended understandings

into theory: and, by suggesting experiments that might be performed, into experiment for verification.

The relationship between theory and experiment is essentially circular. They might be thought of as partners in a very slow conversation carried out over a very long time. And the role of theory is to simplify, to generalise.

Theory from Experiment

The first aspect of theory, theory from experiment, involves pattern finding. Humans look for patterns. Piaget (1955) insists the child develops a view of the world as he/she becomes able to distinguish objects: that is, create constancy between separate perceptions on separate occasions ("object constancy").[14] Pattern finding, the making of one concept from many distinct perceptions, is an intensely human activity. Theories are patterns given widespread credence and accepted as accounting for a part of our experience.

Since, in Popper's characterisation, theories are not provable, they remain only temporarily valid, awaiting disproval. They fall into a category that includes Occam's Razor. The criterion—relative simplicity—of Occam's Razor is no more provable than "randomness" (see Chaitin 1975; Glanville 1977, 1981). To assert something is random is to assert that no pattern has been found, yet. There is no absolute truth in simplicity: rather, there is convenience, coherence and consistency.[15] Occam's criterion can be neither proved (it is a matter of taste) nor properly tested (although it has intuitive validity and we like it to hold—it is the means by which, for example, Newton's universe is subsumed in Einstein's).

Why do we want to simplify?

To make the "continuum" of our experience de-finite, handleable within limited (finite) resources (see Ashby 1964; Glanville 1994, 1997a).

If we did not simplify by (constantly) making (constant) objects constant, we would never be able to recognise them: nor would we "cognise" "perceptions" as being of "objects". The world we lived in would have no object, and we would not be able to conceive, let alone speak, of our own I's. To "cognise" would be beyond inconceivable. We would live in the continuum, the void: about which we cannot speak, for to speak is to distinguish and make objects. Nirvana. In a word.

Nor could we generalise, finding similarities in behaviour and learning from repetition so we can venture the belief that because some (observed) behaviour of some object has always held, it will always hold (see footnotes 7 and 9)!

So strongly do we believe in such simplification that, when we find discrepancies, we explain them away as errors, rather than a demonstration that simplification necessarily omits something. By this device, we maintain our theories.

Theory formalises the significance and necessity of pattern. Pattern gives us objects and recognisable behaviours, allowing us to predict, and risk living by our predictions.[16]

[...]

Theory from Theory (from Experiment)

The second type of theory is the examination of concepts to clarify (hence, develop) these concepts further, reflecting the extended understandings back on the theory and suggesting experiments to be carried out.

While science thought itself essentially empirical, there was always a theoretical area. In some accounts, mathematics (or logic—Russell and Whitehead 1927) is the queen of the sciences. Theoretical science abounds today. For instance, particle physics inhabits a universe of theoretical discourse and is essentially theory driven.[17] Sometimes such areas return to experiment, but not always. Science depends not only on theory based in collecting and organising evidence (simplifying it to form patterns), but also on theory based in examining the consequences of that evidence and its simplification through the logical examination of, for instance, both a particular pattern, and patterns in general.[18]

In building theory from (and of) theory, we use the same devices we use to build theory: simplification, pattern finding. As well as the objects we have found, we treat relations and the patterns they are held to pertain to also as objects. Using the devices of theory on theory, we act self-referentially: and self-reference is, necessarily, circular. We make theory about theory just as we make theory: we find the pattern of pattern.

[...]

This circular process is, I argue, a design process: of continuous modification and unification, the inclusion of more and more in a coherent whole; occasional restart, extension and revolution; the increase in range and of simplification ("Less is More").

[...]

Part II: the World of Design

Reflex

I contend what I described in Part I of this paper is design, and is design at many levels. And, therefore, (scientific) research is a form of design—a specifically restricted form. If this is so, it is inappropriate to require design to be "scientific": for scientific research is a subset (a restricted form) of design, and we do not generally require the set of a subset to act as the sub subset to that subset any more than we require the basement of the building is its attic.

That (scientific) research is a hidden branch of design leads to peculiarities! It is strange an area for so long claiming the uncovering of truth as its purpose itself seems dishonest about what it does and how it does it.

To indicate (scientific) research is a variety of design as forcefully as possible, I shall explain what I mean by "design", reminding the reader how the qualities of that characterisation are found in my earlier description of (scientific) research.[19]

What is Design

There have been many answer to the question of what design is. The characterisation that is used in this paper concentrates on design as a means of exercising our creativity.[20]

Recapitulating, design is a word used in several ways which has, in English, the form both of noun and verb. In this paper, design is mainly thought of as a verb, indicating action. Central to the act of design is circularity. Here, in my view, creativity enters (Glanville 1980a, 1995a), which is at the centre of my interest. Other aspects (e.g., solving a stated problem), although often understood as crucial, are not, I maintain, central to the study of the design act, no matter how important. Problem solving is its own discipline. I am happy to leave it to those interested.[21]

I characterise design (after Pask 1969) as a conversation, usually held via a medium such a paper and pencil, with an other (either an "actual" other or oneself acting as an other) as the conversational partner. The word conversation is used in a recognisable and everyday manner.[22] (Pask eventually developed notions of the conversation into a highly refined technical theory of sophistication and some difficulty Pask 1975, 1976; Glanville 1993, 1998b).

Design-as-conversation will be familiar from the doodle on the back from the envelope upwards. I believe the value of the doodle is an instance of creativity firing the doodler's enthusiasm, personal research and commitment.

Creativity may also be found elsewhere. But this circular process is certainly one in which novelty—a distinguishing feature of design and so typical of creativity—can be generated.[23]

Design and Research

(Scientific) research (whether experiment or theory) is a design activity. We design experiments, but we also act as designers in how we act in these experiments. We design the experiences and objects we find through experiment by finding commonalities (simplification): and we design how we assemble them into patterns (explanatory principles, theories). Looking at these patterns, we make further patterns from them—the theories of our theories. Thus, in doing science, we learn.

The manner in which we do this is circular—conversational (in Pask's sense): we act iteratively, until reaching self-reinforcing stability or misfit. We test, until we arrive at something satisfying our desires—for stability/recognisability/repeatibility/etc. Thus we arrive at our understandings. We test and test again, repeat with refinement and extend, and when driving to extremes we find our patterns no longer hold, we re-jig them or start again from scratch. We adumbrate the special within the more general, coming to resting points where we say (as in design) "this is ok, I can get no further just now".

It is we who do it: we act. The role of observer-as-participant, in making knowledge, abstracting it to theory, theorising about theory; and in constructing the way we obtain this knowledge, then obtaining it accordingly, is central/essential/unavoidable/inevitable and completely desirable. Without the active participation of this actor, there would be nothing that we would know. At every step, in every action, the observer/participant is actively designing. There is nothing passive, automatic or without person (agent, scientist, designer), here.

No matter how regrettable or distasteful this may appear to traditional scientists and others drilled in the convention (the distortion) of presentation by which science puts forward its discoveries and the claims it makes for them, it is a consequence of this examination of how we do science and what we do with what we learn from doing it.

(Scientific) research is a branch of design, in which the designer is central, and through which we construct the world of (and according to) the scientific knowledge we design.

So the act of design, as we understand and value it, has much to offer as an example of how science and scientific research might be in a new era: an era that designer-readers will recognise as their contemporary paradigm and which is how scientists, when we talk to them, recognise and characterise their own activity. Design, being the more general case, satisfies Occam's razor for simplicity: as Einstein is to Newton, design is to science and scientific research.

Conclusion: Research and Design

There are differences between design and (scientific) research: otherwise they would be indistinguishable and we would only need one word (Glanville 1980b). The differences, traditionally emphasised, are not my concern. My intention has been to show that (scientific) research, as it is and must be practised, is properly considered a branch of design: (scientific) research is a subset of design, not the other way round. This is the reason for the potted history in the Prologue. We, who are interested in design and in researching into it, are still inclined to insist we should prosecute our research according to the old and no longer sustainable view of (scientific) research: which view removes from design—and from how we consider and present it—which makes it central, important and valuable, exactly that which characterises it. Even while scientists come to realise their creative involvement in their processes.

We, in design research, should redress this imbalance, indicating the primacy and centrality of design both as an object of study and a means of carrying out that study; insisting on the impropriety of demands that design perform according to criteria of (scientific) research when design is that which encapsulates and embodies this. (Scientific) research should be judged by design criteria, not the other way round. We need to learn to believe in design, to live this, no longer apologising, but refusing to downplay what we do, kowtowing to an old and falsely elevated view.

We should not let the misrepresentations of (scientific) research be forced on us as an insensitive straitjacket. This does not mean we should not glory in the successes (and beauty) of (scientific) research. We should learn from what it offers us, including the lessons of this paper. There are qualities essential (and all too often forgotten) in design which are remembered and given primacy in (scientific) research, such as rigour, honesty, clarification and testing, and the relative strength of argument over assertion. Especially now, when design researchers are again asking about the benefits available from other disciplines, we should look for disciplines that study circularity and the included observer-participant for the insights they may afford us into the operation and consequences of those processes in our research, and what that might mean to us. That is, disciplines that are, at their base, sympathetic to design. Otherwise we forsake our primacy and dance to the wrong tune played by the wrong fiddler, who scarcely believes in the tune any more but who will, nevertheless, call the tune when we ask him to because to do so retains his primacy.

Design is the key to research. Research has to be designed. Considering design carefully (making theory from or even researching it) can reveal how better to act, do research—to design research. And how better to acknowledge design in research: as a way of understanding, acting, looking, searching.

But design should be studied on design's terms. For, design is the form, the basis. And research is a design act. Perhaps that is why it is beautiful?

Design's Secret Partner in Research.

As it happens, there is one subject that is concerned with the philosophical, psychological and mechanical examination of just these issues: cybernetics.

Over the last thirty years, and visible largely through application in other areas, it has (in the form of "second order cybernetics" or the "cybernetics of cybernetics", the "new cybernetics") explored the nature of circular systems and those actions in which the observer (in the most general sense) is a participant. Cybernetics has elucidated conversation, creativity and the invention of the new; multiple viewpoints and their implications for their objects of attention; self-generation and "the emergence" of stability; post rationalisation, representation and experience; constructivism; and distinction drawing and the theory of boundaries.

In this, cybernetics has been explicitly concerned with the qualities we have found to invest research and which are designerly.

This recent manifestation of cybernetics is not to be confused with that for which such large and absurd claims were made at much the same time that the early and determinist ("scientific") approaches in design research were being pushed as the powerful way forward for design. It is a much gentler and more introspective subject, although its approach can be clearly derived from the original (Glanville 1987b, 1998a; von Foerster 1974).

Given this similarity of concern and of formation, it is no surprise that, over these last 30 years, cybernetics has learnt much from design, nor that many of those most intimately involved in the development of this new cybernetics have come from or been closely involved with design.[24]

It is, in my biassed opinion, time that design redressed the balance and examined its Secret Partner in Research, the subject that, learning much from design, has clarified our understandings of designerly qualities. I hope to undertake this in a general manner in a later paper, but, for the meanwhile, the reader is referred to Glanville 1997b, 1998a.

from: Glanville, Ranulph (1999) "Re-searching Design and Designing Research" in: Design Issues 15:2 (summer, 1999), 80-91, © by the Massachusetts Institute of Technology

NOTES

1 This paper was written while Visiting Fellow at the School of Design, Hong Kong Polytechnic University: thanks to them for that most valuable resource, time.
2 This paper is developed from my earlier paper, "Why Design Research" (Glanville 1980).
3 I accept my account may appear inaccurate and oversimplified. But I believe the argument is worth making.
4 It also needed theory, for similar reasons. This need continues today, with the consequent import of endless new theoretical structures from outside design itself.
5 The reader, anxious to know how I use the word design may look at the section "What is Design".
6 I lived in this intellectual environment and believed its simplifications. My (student) sketch books are full of Venn diagrams and directed graphs, rather than sketches of sensitive corners of proposed buildings. I had second chances to study, through teaching and through higher degrees. Otherwise I might still think this way.
7 I prefer the word knowing to knowledge, because knowing requires an agent to know whereas knowledge appears to be knower-free. But, in this paper, I use knowledge to reduce pedantry: please remember, however, that it needs a knower.
8 This is not the place for an argument about the relevance of such a view in a Post-Modern World. Whatever the relative truth of different philosophical positions, science continues to work and to be worked in more or less the manner described here. Arguments about interpretation, personal truth, etc. are close to my heart. As it happens, Cybernetics accommodates these arguments—see the last section.
9 As Wittgenstein (1971) elegantly points out. This is the age-old "Problem" of Induction.
10 I insist investigative actions must have active agents (I am a constructivist), which are, in the case of science, scientists (who are people). See the Black Box model referred to in the next footnote.
11 The theory of the Black Box, which has the added advantage we can never talk about "truth" as a result of using it. See Glanville 1979.
12 That the experimenter is influenced by social factors and epistemological outlook does not reduce his responsibility: he accepts these social factors and acts accordingly.
13 For a rather nice account of the consistent effort not to notice this, and the results of finally realising it cannot be overlooked for ever, see the early parts of James Gleick's account of *Chaos* (Gleick 1987).
14 George Spencer Brown's "Logic of Distinction" (1969) is based on this concept. My first PhD (Glanville 1975) was concerned with how, although we perceive differently we can still believe we see the same "Object".
15 This is the problem facing those wishing to demonstrate absolute "scientific" certainty in, for instance, the non-transmission of BSE, or, more recently, of H5N2 (Hong Kong bird flu) to humans. Popper's point is that science attempts to disprove, so validity is temporary.
16 I will examine the connection between how we think and what design is in another paper. But I believe this account indicates my belief: that design constitutes our way of thinking.
17 The machinations in constructing evidence from photographic "evidence" is astonishing. But not as astonishing as image enhancement creating patterns telling us "truths" in, for instance, space exploration!
18 I do not necessarily mean formal, mathematical logic.

19 I like to believe this holds for all research, because for research to be distinct from assertion requires validation: it is not enough to assemble a few ideas in whatever way we fancy; we must test these ideas (honestly and fairly) for consistency, correspondence with experience (reality), and communion. I shall not pursue this argument here.

20 See Glanville 1980a, 1994, 1995a, 1997b, 1998c. This is my normal characterisation.

21 Some postulate primitive problem solving as a first venture towards design. History is as much a construction as any other account. I do not deny problem solving and design concide. But I insist design takes a space of its own.

22 A conversation is a circular form of communication, in which understandings are exchanged. In a conversation, participants build meanings through the conversational form, rather than trying to communicate a predetermined meaning through coding. In conversation, words do not hold meaning—we do. See Glanville 1995b.

23 Whether the novelty is global, or only to the person designing, at that instant.

24 Particularly, Gordon Pask became a staff member at the Architectural Association, from which school many of his successful doctoral students came, and where many architecture students and teachers learnt quite unwittingly to do second-order cybernetics.

REFERENCES

Ashby, R (1964) "Introductory Remarks at a Panel Discussion" in Mesarovic, M (ed) "Views in General Systems Theory", Chichester: John Wileys and Sons.

Blair, K (1995) "Cubal Grids: Invariable Civilizational Assumptions, Variable Human Values" in Glanville, R and de Zeeuw, G (eds) *Problems of Values and Invariants*, Amsterdam: Thesis Publishers.

Bremmermann, H (1962) "Optimisation Through Evolution and Re-Combination" in Yovits, M, Sawbi, G and Goldstein, G (eds) *Self-Organising Systems*, Washington DC: Spartan Books.

Chaitin, G (1975) "Randomness and Mathematical Proof", *Scientific American*, May.

Feynman, R (1985) *QED: the Strange Theory of Light and Matter*, Princeton: Princeton University Press.

Foerster, H von (1974) *Cybernetics of Cybernetics*, Biological Computer Laboratory Univeristy of Illinois, Champaign–Urbana.

Glanville, R (1975) "A Cybernetic Development of Theories of Epistemology and Observation, with reference to Space and Time, as seen in Architecture" (Ph D Thesis, unpublished) Brunel University, also known as "The Object of Objects, the Point of Points,—or Something about Things").

Glanville, R (1977) "The Nature of Fundamentals, applied to the Fundamentals of Nature" in Klir, G (ed) *Proceedings 1 International Conference on Applied General Systems: Recent Developments & Trends*, New York: Plenum.

Glanville, R (1979) "Inside Every White Box there are two Black Boxes trying to get out", *Behavioural Science* vol 12, no 1, 1982.

Glanville, R (1980a) "The Architecture of the Computable", *Design Studies*, vol 1, No. 4.

Glanville, R (1980b) "The Same is Different" in Zeleny M (ed) *Autopoiesis*, New York: Elsevier.

Glanville, R (1980c) "Why Design Research" in Jacques, R and Powell, J (eds) *Design: Science: Method*, Guilford: Wesbury House.

Glanville, R (1981) "Occam's Adventures in the Black Box" in Lasker, G (ed), *Applied Systems & Cybernetics* vol II, Oxford: Pergamon.

Glanville, R (1987) "The Question of Cybernetics", *Cybernetics, an International Journal*, vol 18, republished in the *General Systems Yearbook*, Society for General Systems Research, Louisville, 1988.

Glanville, R (1993) "Pask: a Slight Primer" in Glanville, R (ed), "Gordon Pask, a Fest-

schrift", *Systems Research* vol 10, no 3.
Glanville, R (1994) "Variety in Design", Systems Research, vol 11, no 3.
Glanville, R (1995a) "Architecture and Computing: a Medium Approach" in Procs 15th Meeting of Association for Computing in Architectural Design in America, University of Washington, Seattle.
Glanville, R (1995b) "Communication without Coding: Cybernetics, Meaning and Language (How Language, becoming a System, Betrays itself)", *Modern Language Notes* Vol 111, no 3.
Glanville, R (1997a) "The Value of being Unmanageable: Variety and Creativity in CyberSpace" Procs of the Conference "Global Village '97", Vienna.
Glanville, R (1997b) "The Value when Cybernetics is Added to CAAD" in Nys, K, Provoost, T, Verbeke, J and Verleye, J (eds) *The Added Value of Computer Aided Architectural Design*, Brussels: Hogeschool voor Wetenschap en Kunst Sint-Lucas.
Glanville, R (1998a) "Cybernetic Realities", Bialystok, Technical University.
Glanville, R (1998b) "Gordon Pask" Luminaries section, web site, www.isss.org
Glanville, R (1998c) "Keeping Faith with the Design in Design Research", Design Research Society conference (ed. Robertson, A), web site www.dmu.ac.uk/ln/4dd/drs9.html
Glanville, R (forthcoming) "Emergence" in review for special issue of *Systems Research*.
Glanville, R, Forey, G and Sengupta, S (forthcoming) "A (Cybernetic) Musing 9: the Language of Science", to be published in *Cybernetics and Human Learning*.
Gleick, J (1987) *Chaos: Making a New Science*, London: Penguin.
Hunston, S (1993) "Evaluation and Ideology in Scientific Writing", in Ghadessy, M (ed) *Register Analysis in Theory and Practice*, London: Pinter Press.
Hyland, K (1997) "Scientific Claims and Community Values: Articulating an Academic Culture", *Language and Communication* Volume 17, no 1.
Kelly, G (1955) "A Theory Of Personality", New York: Norton.
Kuhn, T (1970) "The Nature of Scientific Revolutions" 2nd ed, Chicago: Chicago University Press.
Lakatos, I (1969) "Falsification and the Methodology of Scientific Research Programmes" in Lakatos, I and Musgrave, A (1970) *Criticism and the Growth of Knowledge* Cambridge: Cambridge University Press.
Medawar, P (1963) "Is the Scientific Paper a Fraud?", *The Listener* September 12: 377–8.
Pask, G (1969) "The Architectural Relevance of Cybernetics" *Architectural Design* 9.
Pask, G (1975) *Conversation Theory*, London: Hutchinson.
Pask, G (1976) *Conversation, Cognition and Learning*, New York: Elsevier.
Pask, G, Glanville, R and Robinson, M (1981) *Calculator Saturnalia*, London: Wildwood House and New York: Random House.
Piaget, J (1955) *The Child's Conception of Reality*, New York: Basic Books.
Popper, K (1969) *Conjectures and Refutations* 3rd ed, London: Routledge and Kegan Paul.
Russell, B and Whitehead, A (1927) *Principia Mathematica* 2nd ed, Cambridge: Cambridge University Press.
Spencer Brown, G (1969) *Laws of Form*, London: George Allen and Unwin.
Swanson, G (1997) "Building ISSS Success—One Failure at a Time", Incoming ISSS Presidential Address 1998, web site www.isss.org
Wittgenstein, L (1971) *Tractatus Logico-Philosophicus*, 2nd ed, translated Pears, D and McGuinness, B, London: B. Routledge and Kegan Paul.

Warren Weaver

2 SCIENCE AND COMPLEXITY

Based upon material presented in Chapter 1, "The Scientists Speak," Boni & Gaer Inc.,1947.

SCIENCE has led to a multitude of results that affect men's lives. Some of these results are embodied in mere conveniences of a relatively trivial sort. Many of them, based on science and developed through technology, are essential to the machinery of modern life. Many other results, especially those associated with the biological and medical sciences, are of unquestioned benefit and comfort. Certain aspects of science have profoundly influenced men's ideas and even their ideals. Still other aspects of science are thoroughly awesome.

How can we get a view of the function that science should have in the developing future of man? How can we appreciate what science really is and, equally important, what science is not? It is, of course, possible to discuss the nature of science in general philosophical terms. For some purposes such a discussion is important and necessary, but for the present a more direct approach is desirable. Let us, as a very realistic politician used to say, let us look at the record. Neglecting the older history of science, we shall go back only three and a half centuries and take a broad view that tries to see the main features, and omits minor details. Let us begin with the physical sciences, rather than the biological, for the place of the life sciences in the descriptive scheme will gradually become evident.

Problems of Simplicity

Speaking roughly, it may be said that the seventeenth, eighteenth, and nineteenth centuries formed the period in which physical science learned variables, which brought us the telephone and the radio, the automobile and the airplane, the phonograph and the moving pictures, the turbine and the Diesel engine, and the modern hydroelectric power plant.

The concurrent progress in biology and medicine was also impressive, but that was of a different character. The significant problems of living organisms are seldom those in which one can rigidly maintain constant all but two variables. Living things are more likely to present situations in which a half-dozen, or even several dozen quantities are all varying simultaneously, and in subtly interconnected ways. Often they present situations in which the essentially important quantities are either non-quantitative, or have at any rate eluded identification or measurement up to the moment. Thus biological and medical problems often involve the consideration of a most complexly organized whole. It is not surprising that up to 1900 the life sciences were largely concerned with the necessary preliminary stages in the application of the scientific method-preliminary stages which chiefly involve collection, description, classification, and the observation of concurrent and apparently correlated effects. They had only made the brave beginnings of quantitative theories, and hardly even begun detailed explanations of the physical and chemical mechanisms underlying or making up biological events.

To sum up, physical science before 1900 was largely concerned with two-variable problems of simplicity; whereas the life sciences, in which these problems of simplicity are not so often significant, had not yet become highly quantitative or analytical in character.

Problems of Disorganized Complexity

Subsequent to 1900 and actually earlier, if one includes heroic pioneers such as Josiah Willard Gibbs, the physical sciences developed an attack on nature of an essentially and dramatically new kind. Rather than study problems which involved two variables or at most three or four, some imaginative minds went to the other extreme, and said: "Let us develop analytical methods which can deal with two billion variables." That is to say, the physical scientists, with the mathematicians often in the vanguard, developed powerful techniques of probability theory and of statistical mechanics to deal with what may he called problems of disorganized complexity.

This last phrase calls for explanation. Consider first a simple illustration in order to get the flavor of the idea. The classical dynamics of the nineteenth century was well suited for analyzing and predicting the motion of a single ivory ball as it moves about on a billiard table. In fact, the relationship between positions of the ball and the times at which it reaches these positions forms a typical nineteenth-century problem of simplicity. One can, but with a surprising increase in difficulty, analyze the motion of two or even of three balls on a billiard table. There has been, in fact, considerable study of the mechanics of the standard game of billiards. But, as soon as one tries to analyze the motion of ten or fifteen balls on the table at once, as in pool, the problem becomes unmanageable, not because there is any theoretical difficulty, but just because the actual labor of dealing in specific detail with so many variables turns out to be impracticable.

Imagine, however, a large billiard table with millions of balls rolling over its surface, colliding with one another and with the side rails. The great surprise is that the problem now becomes easier, for the methods of statistical mechanics are applicable. To be sure the detailed history of one special ball can not be traced, but certain important questions can be answered with useful precision, such as: On the average how many balls per second hit a given stretch of rail? On the average how far does a ball move before it is hit by some other ball? On the average how many impacts per second does a ball experience?

Earlier it was stated that the new statistical methods were applicable to problems of disorganized complexity. How does the word "disorganized" apply to the large billiard table with the many balls? It applies because the methods of statistical mechanics are valid only when they are distributed, in their positions and motions, in a helter-skelter, that is to say a disorganized, way. For example, the statistical methods would not apply if someone were to arrange the balls in a row parallel to one side rail of the table, and then start them all moving in precisely parallel paths perpendicular to the row in which they stand. Then the balls would never collide with each other nor with two of the rails, and one would not have a situation of disorganized complexity.

From this illustration it is clear what is meant by a problem of disorganized complexity. It is a problem in which the number of variables is very large, and one in which each of the many variables has a behavior which is individually erratic, or perhaps totally unknown. However, in spite of this helter-skelter, or unknown, behavior of all the individual variables, the system as a whole possesses certain orderly and analyzable average properties.

A wide range of experience comes under the label of disorganized complexity. The method applies with increasing precision when the number of variables increases. It applies with entirely useful precision to the experience of a large telephone exchange, in predicting the average frequency of calls, the probability of overlapping calls of the same number, etc. It makes possible the financial stability of a life insurance company. Although the company can have no knowledge whatsoever concerning the approaching death of any one individual, it has dependable knowledge of the average frequency with which deaths will occur.

This last point is interesting and important. Statistical techniques are not restricted to situations where the scientific theory of the individual events is very well known, as in the billiard example where there is a beautifully precise theory for the impact of one ball on another. This technique can also be applied to situations, like the insurance example, where the individual event is as shrouded in mystery as is the chain of complicated and unpredictable events associated with the accidental death of a healthy man.

The examples of the telephone and insurance companies suggests a whole array of practical applications of statistical techniques based on disorganized complexity. In a sense they are unfortunate examples, for they tend to draw attention away from the more fundamental use which science makes of these new techniques. The motions of the atoms which form all matter, as well as the motions of the stars which form the universe, come under the range of these new techniques. The fundamental laws of heredity are analyzed by them. The laws of thermodynamics, which describe basic and inevitable tendencies of all physical systems, are derived from statistical considerations. The entire structure of modem physics, our present concept of the nature of the physical universe, and of the accessible experimental facts concerning it rest on these statistical concepts. Indeed, the whole question of evidence and the way in which knowledge can be inferred from evidence are now recognized to depend on these same statistical ideas, so that probability notions are essential to any theory of knowledge itself.

Problems of Organized Complexity

This new method of dealing with disorganized complexity, so powerful an advance over the earlier two-variable methods, leaves a great field untouched. One is tempted to oversimplify, and say that scientific methodology went from one extreme to the other—from two variables to an astronomical number—and left untouched a great middle region. The importance of this middle region, moreover, does not depend primarily on the fact that the number of variables involved is moderate—large compared to two, but small compared to the number of atoms in a pinch of salt. The problems in this middle region, in fact, will often involve a considerable number of variables. The really important characteristic of the problems of this middle region, which science has as yet little explored or conquered, lies in the fact that these problems, as contrasted with the disorganized situations with which statistics can cope, show the essential feature of organization. In fact, one can refer to this group of problems as those of organized complexity.

What makes an evening primrose open when it does? Why does salt water fail to satisfy thirst? Why can one particular genetic strain of microorganism synthesize within its minute body certain organic compounds that another strain of the same organism cannot manufacture? Why is one chemical substance a poison when another, whose molecules have just the same atoms but assembled into a mirror-image pattern, is completely harmless? Why does the amount of manganese in the diet affect the maternal instinct of an animal? What is the description of aging in biochemical terms? What meaning is to be assigned to the question:

Is a virus a living organism? What is a gene, and how does the original genetic constitution of a living organism express itself in the developed characteristics of the adult? Do complex protein molecules "know how" to reduplicate their pattern, and is this an essential clue to the problem of reproduction of living creatures? All these are certainly complex problems, but they are not problems of disorganized complexity, to which statistical methods hold the key. They are all problems which involve dealing simultaneously with a sizable number of factors which are interrelated into an organic whole. They are all, in the language here proposed, problems of organized complexity.

On what does the price of wheat depend? This too is a problem of organized complexity. A very substantial number of relevant variables is involved here, and they are all interrelated in a complicated, but nevertheless not in helter-skelter, fashion.

How can currency be wisely and effectively stabilized? To what extent is it safe to depend on the free interplay of such economic forces as supply and demand? To what extent must systems of economic control be employed to prevent the wide swings from prosperity to depression? These are also obviously complex problems, and they too involve analyzing systems which are organic wholes, with their parts in close interrelation.

How can one explain the behavior pattern of an organized group of persons such as a labor union, or a group of manufacturers, or a racial minority? There are clearly many factors involved here, but it is equally obvious that here also something more is needed than the mathematics of averages. With a given total of national resources that can be brought to bear, what tactics and strategy will most promptly win a war, or better: what sacrifices of present selfish interest will most effectively contribute to a stable, decent, and peaceful world?

These problems—and a wide range of similar problems in the biological, medical, psychological, economic, and political sciences—are just too complicated to yield to the old nineteenth-century techniques which were so dramatically successful on two-, three-, or four-variable problems of simplicity. These new problems, moreover, cannot be handled with the statistical techniques so effective in describing average behavior in problems of disorganized complexity.

These new problems, and the future of the world depends on many of them, requires science to make a third great advance, an advance that must be even greater than the nineteenth-century conquest of problems of simplicity or the twentieth-century victory over problems of disorganized complexity. Science must, over the next fifty years, learn to deal with these problems of organized complexity.

Is there any promise on the horizon that this new advance can really be accomplished? There is much general evidence, and there are two recent instances of especially promising evidence. The general evidence consists in the fact that, in the minds of hundreds of scholars all over the world, important, though necessarily minor, progress is already being made on such problems. As never before, the quantitative experimental methods and the mathematical analytical methods of the physical sciences are being applied to the biological, the medical, and even the social sciences. The results are as yet scattered, but they are highly promising. A good illustration from the life sciences can be seen by a comparison of the present situation in cancer research with what it was twenty-five years ago. It is doubtless true that we are only scratching the surface of the cancer problem, but at least there are now some tools to dig with and there have been located some spots beneath which almost surely there is pay-dirt. We know that certain types of cancer can be induced by certain pure chemicals. Something is known of the inheritance of susceptibility to certain types of cancer. Million-volt rays are available, and the even more intense radiations made possible by atomic physics. There are radioactive isotopes, both for basic studies and for treatment. Scientists are tackling the almost incredibly complicated story of the biochemistry of the aging organism. A base of knowledge concerning the normal cell is being established that makes it possible to recognize and analyze the pathological cell. However distant the goal, we are now at last on the road to a successful solution of this great problem.

In addition to the general growing evidence that problems of organized complexity can be successfully treated, there are at least two promising bits of special evidence. Out of the wickedness of war have come two new developments that may well be of major importance in helping science to solve these complex twentieth-century problems.

The first piece of evidence is the wartime development of new types of electronic computing devices. These devices are, in flexibility and capacity, more like a human brain than like the traditional mechanical computing device of the past. They have memories in which vast amounts of information can be stored. They can be "told" to carry out computations of very intricate complexity, and can be left unattended while they go forward automatically with their task. The astounding speed with which they proceed is illustrated by the fact that one small part of such a machine, if set to multiplying two ten-digit numbers, can perform such multiplications some 40,000 times faster than a human operator can say "Jack Robinson." This combination of flexibility, capacity, and speed makes it seem likely that such devices will have a tremendous impact on science. They will make it possible to deal with problems which previously were too complicated, and, more importantly, they will justify and inspire the development of new methods of analysis applicable to these new problems of organized complexity.

The second of the wartime advances is the "mixed-team" approach of operations analysis. These terms require explanation, although they are very familiar to those who were concerned with the application of mathematical methods to military affairs.

As an illustration, consider the overall problem of convoying troops and supplies across the Atlantic. Take into account the number and effectiveness of the naval vessels available, the character of submarine attacks, and a multitude of other factors, including such an imponderable as the dependability of visual watch when men are tired, sick, or bored. Considering a whole mass of factors, some measurable and some elusive, what procedure would lead to the best overall plan, that is, best from the combined point of view of speed, safety, cost, and so on? Should the convoys be large or small, fast or slow? Should they zigzag and expose themselves longer to possible attack, or dash in a speedy straight line? How are they to be organized, what defenses are best, and what organization and instruments should be used for watch and attack?

The attempt to answer such broad problems of tactics, or even broader problems of strategy, was the job during the war of certain groups known as the operations analysis groups. Inaugurated with brilliance by the British, the procedure was taken over by this country, and applied with special success in the Navy's anti-submarine campaign and in the Army Air Forces. These operations analysis groups were, moreover, what may be called mixed teams. Although mathematicians, physicists, and engineers were essential, the best of the groups also contained physiologists, biochemists, psychologists, and a variety of representatives of other fields of the biochemical and social sciences. Among the outstanding members of English mixed teams, for example, were an endocrinologist and an X-ray crystallographer. Under the pressure of war, these mixed teams pooled their resources and focused all their different insights on the common problems. It was found, in spite of the modern tendencies toward intense scientific specialization, that members of such diverse groups could work together and could form a unit which was much greater than the mere sum of its parts. It was shown that these groups could tackle certain problems of organized complexity, and get useful answers.

It is tempting to forecast that the great advances that science can and must achieve in the next fifty years will be largely contributed to by voluntary mixed teams, somewhat similar to the operations analysis groups of war days, their activities made effective by the use of large, flexible, and highspeed computing machines. However, it cannot be assumed that this will be the exclusive pattern for future scientific work, for the atmosphere of complete intellectual freedom is essential to science. There will always, and properly, remain those scientists for whom intellectual freedom is necessarily a private affair. Such men must, and should, work alone. Certain deep and imaginative achievements are probably won only in such a way. Variety is, moreover, a proud characteristic of the American way of doing things. Competition between all sorts of methods is good. So there is no intention here to picture a future in which all scientists are organized into set patterns of activity. Not at all. It is merely suggested that some scientists will seek and develop for themselves new kinds of collaborative arrangements; that these groups will have members drawn from essentially all fields of science; and that these new ways of working, effectively instrumented by huge computers, will contribute greatly to the advance which the next half century will surely achieve in handling the complex, but essentially organic, problems of the biological and social sciences.

The Boundaries of Science

Let us return now to our original questions. What is science? What is not science? What may be expected from science?

Science clearly is a way of solving problems—not all problems, but a large class of important and practical ones. The problems with which science can deal are those in which the predominant factors are subject to the basic laws of logic, and are for the most part measurable. Science is a way of organizing reproducible knowledge about such problems; of focusing and disciplining imagination; of weighing evidence; of deciding what is relevant and what is not; of impartially testing hypotheses; of ruthlessly discarding data that prove to be inaccurate or inadequate; of finding, interpreting, and facing facts, and of making the facts of nature the servants of man.

The essence of science is not to be found in its outward appearance, in its physical manifestations; it is to be found in its inner spirit. That austere but exciting technique of inquiry known as the scientific method is what is important about science. This scientific method requires of its practitioners high standards of personal honesty, open-mindedness, focused vision, and love of the truth. These are solid virtues, but science has no exclusive lien on them. The poet has these virtues also, and often turns them to higher uses.

Science has made notable progress in its great task of solving logical and quantitative problems. Indeed, the successes have been so numerous and striking, and the failures have been so seldom publicized, that the average man has inevitably come to believe that science is just about the most spectacularly successful enterprise man ever launched. The fact is, of course, that this conclusion is largely justified.

Impressive as the progress has been, science has by no means worked itself out of a job. It is soberly true that science has, to date, succeeded in solving a bewildering number of relatively easy problems, whereas the hard problems, and the ones which perhaps promise most for man's future, lie ahead.

We must, therefore, stop thinking of science in terms of its spectacular successes in solving problems of simplicity. This means, among other things, that we must stop thinking of science in terms of gadgetry. Above all, science must not be thought of as a modern improved black magic capable of accomplishing anything and everything.

Every informed scientist, I think, is confident that science is capable of tremendous further contributions to human welfare. It can continue to go forward in its triumphant march against physical nature, learning new laws, acquiring new power of forecast and control, making new material things for man to use and enjoy. Science can also make further brilliant contributions to our understanding of animate nature, giving men new health and vigor, longer and more effective lives, and a wiser understanding of human behavior. Indeed, I think most informed scientists go even further and expect that the precise, objective, and analytical techniques of science will find useful application in limited areas of the social and political disciplines.

There are even broader claims which can be made for science and the scientific method. As an essential part of his characteristic procedure, the scientist insists on precise definition of terms and clear characterization of his problem. It is easier, of course, to define terms accurately in scientific fields than in many other areas. It remains true, however, that science is an almost overwhelming illustration of the effectiveness of a well-defined and accepted language, a common set of ideas, a common tradition. The way in which this universality has succeeded in cutting across barriers of time and space, across political and cultural boundaries, is highly significant. Perhaps better than in any other intellectual enterprise of man, science has solved the problem of communicating ideas, and has demonstrated the world-wide cooperation and community of interest which then inevitably results.

Yes, science is a powerful tool, and it has an impressive record. But the humble and wise scientist does not expect or hope that science can do everything. He remembers that science teaches respect for special competence, and he does not believe that every social, economic, or political emergency would be automatically dissolved if "the scientists" were only put into control. He does not—with a few aberrant exceptions—expect science to furnish a code of morals, or a basis for esthetics. He does not expect science to furnish the yardstick for measuring, nor the motor for controlling, man's love of beauty and truth, his sense of value, or his convictions of faith. There are rich and essential parts of human life which are alogical, which are immaterial and non-quantitative in character, and which cannot be seen under the microscope, weighed with the balance, nor caught by the most sensitive microphone.

If science deals with quantitative problems of a purely logical character, if science has no recognition of or concern for value or purpose, how can modern scientific man achieve a balanced good life, in which logic is the companion of beauty, and efficiency is the partner of virtue:

In one sense the answer is very simple: our morals must catch up with our machinery. To state the necessity, however, is not to achieve it. The great gap, which lies so forebodingly between our power and our capacity to use power wisely, can only be bridged by a vast combination of efforts. Knowledge of individual and group behavior must be improved. Communication must be improved between peoples of different languages and cultures, as well as between all the varied interests which use the same language, but often with such dangerously differing connotations. A revolutionary advance must be made in our understanding of economic and political factors. Willingness to sacrifice selfish short-term interests, either personal or national, in order to bring about long-term improvement for all must be developed.

None of these advances can be won unless men understand what science really is; all progress must be accomplished in a world in which modern science is an inescapable, ever-expanding influence.

from: American Scientist, 36: 536 (1948), based upon material presented in Chapter 1 "The Scientists Speak," Boni & Gaer Inc., 1947

Herbert A. Simon

3 SOCIAL PLANNING: DESIGNING THE EVOLVING ARTIFACT

In chapter 5 I surveyed some of the modern tools of design that are used by planners and artificers. Even before most of these tools were available to them, ambitious planners often took whole societies and their environments as systems to be refashioned. Some recorded their utopias in books—Plato, Sir Thomas More, Marx. Others sought to realize their plans by social revolution in America, France, Russia, China. Many or most of the large-scale designs have centered on political and economic arrangements, but others have focused on the physical environment—river development plans, for example, reaching from ancient Egypt to the Tennessee Valley to the Indus and back to today's Nile.

As we look back on such design efforts and their implementation, and as we contemplate the tasks of design that are posed in the world today, our feelings are very mixed. We are energized by the great power our technological knowledge bestows on us. We are intimidated by the magnitude of the problems it creates or alerts us to. We are sobered by the very limited success—and sometimes disastrous failure—of past efforts to design on the scale of whole societies. We ask, "If we can go to the Moon, why can't we . . . ?"—not expecting an answer, for we know that going to the Moon was a simple task indeed, compared with some others we have set for ourselves, such as creating a humane society or a peaceful world. Wherein lies the difference?

Going to the Moon was a complex matter along only one dimension: it challenged our technological capabilities. Though it was no mean accomplishment, it was achieved in an exceedingly cooperative environment, employing a single new organization, NASA, that was charged with a single, highly operational goal. With enormous resources provided to it, and operating through well-developed market mechanisms, that organization could draw on the production capabilities and technological sophistication of our whole society. Although several potential side effects of the activity (notably its international political and military significance, and the possibility of technological spinoffs) played a major role in motivating the project, they did not have to enter much into the thoughts of the planners once the goal of placing human beings on the Moon had been set. Moreover these by-product benefits and costs are not what we mean when we say the project was a success. It was a success because people walked on the surface of the Moon. Nor did anyone anticipate what turned out to be one of the more important consequences of these voyages: the vivid new perspective we gained of our place in the universe when we first viewed our own pale, fragile planet from space.

Consider now a quite different example of human design. Some twenty years ago we celebrated the 200th birthday of our nation, and about a decade ago we celebrated the 200th anniversary of the framing of its political constitution. Almost all of us in the free world regard that document as an impressive example of success in human planning. We regard it as a success because the Constitution, much modified and much interpreted, still survives as the framework for our political institutions, and because the society that operates within its framework has provided most of us with a broad range of freedoms and a high level of material comfort.

Both achievements—the voyages to the Moon and the survival of the American Constitution—are triumphs of bounded rationality. A necessary, though not a sufficient, condition of their success was that they were evaluated against limited objectives. I have already argued that point with respect to NASA. As to the founding fathers, it is instructive to examine their own views of their goals, reflected in *The Federalist* and the surviving records of the constitutional convention.[1] What is striking about these documents is their practical sense and the awareness they exude of the limits of foresight about large human affairs. Most of the framers of the Constitution accepted very restricted objectives for their artifact—principally the preservation of freedom in an orderly society. Moreover they did not postulate a new man to be produced by the new institutions but accepted as one of their design constraints the psychological characteristics of men and women as they knew them, their selfishness as well as their common sense. In their own cautious words (*The Federalist*, no. 55), "As there is a degree of depravity in mankind which requires a certain degree of circumspection and distrust, so there are other qualities in human nature which justify a certain portion of esteem and confidence."

These examples illustrate some of the characteristics and complexities of designing artifacts on a societal scale. The success of planning on such a scale may call for modesty and restraint in setting the design objectives and drastic simplification of the real-world situation in representing it for purposes of the design process. Even with restraint and simplification difficult obstacles must usually be surmounted to reach the design objetives. The obstacles and some of the techniques for overcoming them provide the main subject of this chapter.

Our first topic will be problem representation; our second, ways of accommodating to the inadequacies that can be expected in data; our third, how the nature of the client affects planning; our fourth, limits on the planner's time and attention; and our fifth, the ambiguity and conflict of goals in societal planning. These topics, which can be viewed as a budget of obstacles or alternatively as a budget of planning requirements, will suggest to us some additions to the curriculum in design outlined in the last chapter.

[...]

Society as the Client

It may seem obvious that all ambiguities should be resolved by identifying the client with the whole society. That would be a clear-cut solution in a world without conflict of interest or uncertainty in professional judgment. But when conflict and uncertainty are present, it is a solution that abdicates organized social control over professionals and leaves it to them to define social goals and priorities. If some measure of control is to be maintained, the institutions of the society must share with the professional the redefinition of the goals of design.

The client seeks to control professionals not only by defining their goals of design but also by reacting to the plans they propose. It is well known that physicians' patients fail to take much of the medicine that is prescribed for them. Society as client is no more docile than are medical patients. In any planning whose implementation involves a pattern of human behavior, that behavior must be motivated. Knowledge that "it is for your own good" seldom provides adequate motivation.

The members of an organization or a society for whom plans are made are not passive instruments, but are themselves designers who are seeking to use the system to further their own goals. Organization theory deal with this motivational question by examining organizations of terms of the balance between the inducements that are provided to members to perform their organizational roles and the contributions that the members thereby provide to the achievement of organizational goals.[2]

A not dissimilar representation of the social planning process views it as a game between the planners and those whose behavior they seek to influence. The planners make their move (i.e., implement their design), and those who are affected by it then alter their own behavior to achieve their goals in the changed environment. The gaming aspects of social planning are particularly evident in the domain of economic stabilization policies, where the adaptive response of firms and consumers to monetary and fiscal policies may largely neutralize or negate those policies. The claims of monetarists, and especially of the "rational expectations" theorists, that government is helpless to influence employment levels by using the standard Keynesian tools of monetary and fiscal policy and that attempts to reduce unemployment can only cause inflation, are based on the assumption that public responses to these measures will be strongly and rapidly adaptive.

Except for economics it is still relatively rare for social planning and policy discussions to include in any systematic way the possible "gaming" responses to plans. For example, until quite recently it was common to design new urban transit facilities without envisioning the possible relocations of population within the urban area that would be produced by the new facilities themselves. Yet such effects have been known and observed for half a century. Social planning techniques need to be expanded to encompass them routinely.

Organizations in Social Design

In introducing the subject of social design, I used the Constitution of the United States as an example. Configuring organizations, whether business corporations, governmental organizations, voluntary societies or others, is one of society's most important design tasks. If we human beings were isolated monads—small, hermetically sealed particles that had no mutual relations except occasional elastic collisions—we would not have to concern ourselves with the design of organizations. But, contrary to libertarian rhetoric, we are not monads. From birth until death, our ability to reach our goals, even to survive, is tightly linked to our social interactions with others in our society.

The rules imposed upon us by organizations—the organizations that employ us and the organizations that govern us—restrict our liberties in a variety of ways. But these same organizations provide us with opportunities for reaching goals and attaining freedoms that we could not even imagine reaching by individual effort. For example, almost everyone who will read these lines has an income that is astronomical by comparison with the world average. If we were to assign a single cause to our good fortune, we would have to attribute it to being born in the right place at the right time: in a society that is able to maintain order (through public organizations), to produce efficiently (largely through business organizations), and to maintain the infrastructure required for high production (again largely through public organizations). We have even discovered, in our society and a modest number of others, how to design organizations, business and governmental, that do not interfere egregiously with our freedoms, including those of speech and thought.

This is not the place to enter into a long disquisition on organizational design, private and public, which has a large literature of its own.[3] But one can hardly pass by governments and business firms in complete silence in a chapter on the design of social structures. A society's organizations are matters not only of specialized professional concern but of broad public concern.

Today, organizations, and especially governmental organizations, have an exceedingly bad press in our society. The terms "politician" and "bureaucrat" are not used as descriptors but as pejoratives. While the events in Oklahoma City surely did not evoke public approval, the general horrified reaction was not to the anti-governmental attitudes that the bombing expressed but to the killings. There is more than a little anarchism (usually phrased as libertarianism) in the current American credo (and for that matter, in our credo since the time of the Founding Fathers).

Organizational design, then, is a matter for urgent attention in any curriculum on social design. Organizations are exceedingly complex systems that share many properties with other complex systems, for example, their typically hierarchical structure. Questions of organizational design will reappear, from time to time, as part of the discussion of complex-systems in chapters 7 and 8, below, and especially in connection with the use of hierarchy and "near decomposability" as a basis for specialization.

Time and Space Horizons for Design

Each of us sits in a long dark hall within a circle of light cast by a small lamp. The lamplight penetrates a few feet up and down the hall, then rapidly attenuates, diluted by the vast darkness of future and past that surrounds it. We are curious about that darkness. We consult soothsayers and forecasters of the economy and the weather, but we also search backward for our "roots." Some years ago I conducted such a search in the Rheinland villages near Mainz where my paternal ancestors had lived. I found records of grandparents readily enough, and even of great-grandparents and beyond. But before I had gone far—scarcely back to the 18th century—I came to the edge of the circle of light. Darkness closed in again in the little towns of Ebersheim, Woerstadt, and Partenheim, and I could see no farther back.

History, archaeology, geology, and astronomy provide us with narrow beams that penetrate immense distances down the hallway of the past but illuminate it only fitfully—a statesman or philosopher here, a battle there, some hominoid bones buried with pieces of chipped stone, fossils embedded in ancient rock, rumors of a great explosion. We read about the past with immense interest. A few spots caught by the beams take on a vividness and immediacy that capture, for a moment, our attention and our hearts—some Greek warriors camped before Troy, a man on a cross, the painted figure of a deer glimpsed by flickering torchlight on the wall of a limestone cave. But mostly the figures are shadowy, and our attention shifts back to the present.

The light dims even more rapidly in the opposite direction, toward the future. Although we are titillated by Sunday Supplement descriptions of a cooling Sun, it is our own mortality, just a few years away, and not the Earth's, with which we are preoccupied. We can empathize with parents and grandparents whom we have known, or of whom we have had firsthand accounts, and in the opposite direction with children and grandchildren. But beyond that circle, our concern is more curious and intellectual than emotional. We even find it difficult to define which distant events are the triumphs and which the catastrophes, who the heroes and who the villains.

Discounting the Future

Thus the events and prospective events that enter into our value systems are all dated, and the importance we attach to them generally drops off sharply with their distance in time. For the creatures of bounded rationality that we are, this is fortunate. If our decisions depended equally upon their remote and their proximate consequences, we could never act but would be forever lost in thought. By applying a heavy discount factor to events, attenuating them with their remoteness in time and space, we reduce our problems of choice to a size commensurate with our limited computing capabilities. We guarantee that, when we integrate outcomes over the future and the world, the integral will converge.

Sociobiologists, in their analyses of egoism and altruism, undertake to explain how the forces of evolution would necessarily produce organisms more protective of their offspring and their kin than of unrelated creatures. This evolutionary account does not explain, however, why the concern tends to be so myopic with respect to the future. At least one part of the explanation is that we are unable to think coherently about the remote future, and particularly about the distant consequences of our actions. Our myopia is not adaptive, but symptomatic of the limits of our adaptability. It is one of the constraints on adaptation belonging to the inner environment.

The economist expresses this discounting of the future by a rate of interest. To find the present value of a future dollar, he applies, backwards, a compound discount rate that shrinks the dollar by a fixed percentage for each step from the present. Even a moderate rate of interest can make the dollars of the next century look quite inconsequential for our present decisions. There is a vast literature seeking to explain, none too convincingly, what determines the time rate of discount used by savers. (In modern times it has hovered remarkably steadily around 3 percent per annum, after appropriate adjustment for risk and inflation.) There is also a considerable literature seeking to determine what the social rate of interest *should* be—what the rate of exchange should be between the welfare of this generation and the welfare of its descendants.

The rate of interest should not be confused with another factor that discounts the importance of the future with respect to the present. Even we are aware of certain unfavorable events that will occur in the distant future. there may be nothing to be done about them today. If we knew that the wheat harvest was going to fail in the year 2020, we would be ill-advised to store wheat now. Our unconcern with a distant future is not merely a failure of empathy but a recognition that (1) we shall probably not be able to foresee and calculate the consequences of our actions for more than short distances into the future and (2) those consequences will in any case be diffuse rather than specific.

The important decisions we make about the future are primarily the decisions about spending and saving—about how we shall allocate our production between present and future satisfactions. And in saving, we count flexibility among the important attributes of the objects of our investment, because flexibility insures the value of those investments against the events that will surely occur but which we cannot predict. It will (or should) bias our investments in the direction of structures that can be shifted from one use to another, and to knowledge that is fundamental enough not soon to be outmoded—knowledge that may itself provide a basis for continuing adaptation to the changing environment.

The Change in Time Perspective

One of the noteworthy characteristics of our century is the shift that appears to be taking place, especially in the industrialized world, in our time prespectives. For example, embedded in the energy-environment problem that confronts us today, we can see three almost independent aspects. The first is our immediate dependence on petroleum, which we must reduce to protect ourselves from political blackmail and to achieve a balance of international payments. The second is the prospect of the exhaustion of oil and gas supplies, a problem that must be solved within about a generation, mostly the use of coal and nuclear energy. The third is the joint problem of the exhaustion of fossil fuels and the impact of their combustion on the climate. The time scale of this third problem is a century or so.

What is remarkable in our age, and relatively novel I believe, is the amount of attention we pay to the third problem. Perhaps it is just that we have all three confused in our minds and have not sorted them out to the point where we can think about the more pressing ones without concern for the other. But I do not think that is the reason. I believe there has been a genuine downward shift in the social interest rate we apply to discount events that are remote in time and space.

There are some obvious reasons for our new concern with matters that are remote in time and space. Among these are the relatively new facts of instantaneous worldwide communication and rapid air transportation. Consequent on these is the continually increasing economic and military interdependence of all the nations. More subtle than either of these causes is the progress of human knowledge, especially of science. I have already commented on the way in which archaeology, geology, anthropology, and cosmology have lengthened our perspectives. But in addition new laboratory technologies have vastly increased our ability to detect and assess small and indirect effects of our actions. Oscar Wilde once claimed that there were no fogs on the River Thames until Turner, by painting them, revealed them to the residents of London. In the same way our atmosphere contained no noxious substances in quantities of a few parts per million until chromatography and other sensitive analytic techniques showed their presence and measured them. DDT was an entirely beneficent insecticide until we detected its presence in falcons' eggs and in fish. If eating the apple revealed to us the nature of good and evil, modern analytic tools have taught us how to detect good and evil in minute amounts and at immense distances in time and space.

It may be objected that there has been no such lengthening of social time perspectives as I have claimed. What perspective can be longer than the eternity of life after death that is so central to Christian thought, or longer than the repeated reincarnations of Eastern religions? But the attitudes toward the future engendered by those beliefs are very different from the ones I have been discussing. The future with which the Christian is concerned is his own future in the light of his current conduct. There is nothing in the belief in an afterlife or a reincarnation that calls attention to the future consequences for this world of one's present actions. Nor do I find in those religious beliefs anything resembling the contemporary concern for the fragility of the environment on which human life depends or for the power of human actions to make that environment more or less habitable in the future. It does appear therefore that there has been a genuine shift in our orientation to time and a significant lengthening in time perspectives.

Defining Progress

As the web of cause and effect is woven tighter, we put severe loads upon our planning and decision-making procedures to deal with these remote effects. There is a continuing race between the part of our new science and knowledge that enables us to see more distant views and the part that enables us to deal with what we see. And if we live in a time that is sometimes pessimistic about technology, it is because we have learned to look farther than our arms can reach.

Defining what is meant by progress in human societies is not easy. Increasing success in meeting basic human needs for food, shelter, and health is one kind of definition that most people would agree upon. Another would be an average increase in human happiness. With the advance of productive technology, we can claim that there has been major progress by the first criterion; but what has been said in chapter 2 about changing aspiration levels would lead us to doubt whether progress is possible if we use the second criterion, human happiness, to measure it. There is no reason to suppose that a modern industrial society is more conducive to human happiness than the simpler, if more austere, societies that preceded it. On the other hand, there seems to be little empirical basis for the nostalgia that is sometimes expressed for an imagined (and imaginary) happier or more humane past.

A third way of measuring progress is in terms of intentions rather than outcomes—what might be called moral progress. Moral progress has always been associated with the capacity to respond to universal values—to grant equal weight to the needs and claims of all mankind, present and future. It can be argued that the growth of knowledge of the kinds I have been describing represents such moral progress.

But we should not be hasty in our evaluation of the consequences of lengthening perspectives in space or time. The present century is not lacking in horrible examples of man's inhumanity to man. We must be [...] also to the possibility that rationality applied to a broader domain will simply be a more calculatedly rational selfishness than the impulsive selfishness of the past.

The Management of Attention

From a pragmatic standpoint we are concerned with the future because securing a satisfactory future may require actions in the present. Any interest in the future that goes beyond this call for present action has to be charged to pure curiosity. It belongs to our recreational rather than our working day. Thus our present concern for the short-run energy problem is quite different from our concern for the long-run problem or even the middle-run problem. The actions we have to take today if we are to improve the short-run situation, are largely actions that will reduce our use of energy—there are only modest prospects of a substantial short-run increase in supply. The actions we have to take with respect to the middle-run problem are largely actions on a large scale toward the development and exploitation of some mix of technologies for the conversion of coal mining of oil sands and shales, and safe nuclear fission or fusion. The principal actions we can take now with respect to the long-range energy problem are primarily knowledge-acquiring actions—research programs to develop nuclear fusion and solar technologies and to gain a deeper understanding of the environmental consequences of all the alternatives.

The energy problem is rather typical in this respect of large-scale design problems. In addition to the things we can do to produce immediate consequences, we must anticipate the time lags involved in developing new capital plant and the even greater time lags involved in developing the body of technology and other knowledge that we will need in the more distant future. Attention of the decision-making bodies has to be allocated correspondingly.

It is a commonplace organizational phenomenon that attending to the needs of the moment—putting out fires—takes precedence over attending to the needs for new capital investment or new knowledge. The more crowded the total agenda and the more frequently emergencies arise, the more likely it is that the middle-range and long range decisions will be neglected. In formal organizations a remedy is often sought for this condition by creating planning groups that are insulated in one way or another from the momentary pressures upon the organization. Planning units face two hazards. On the one hand, and especially if they are competently staffed, they may be consulted more and more frequently for help on immediate problems until they are sucked into the operating organization and can no longer perform their planning functions. If they are sufficiently well sealed off from the rest of the organization to prevent this from happening, then they may find the reverse channel blocked—they may be unable to influence decisions in the operating organization. There is no simple or automatic way to remove these difficulties once and for all. They require repeated attention from the organization's leadership.

Designing without Final Goals

To speak of planning without goals may strike one as a contradiction in terms.[4] It seems "obvious" that the very concept of rationality implies goals at which thought and action are aimed. How can we evaluate a design unless we have well-defined criteria against which to judge it, and how can the design process itself proceed without such criteria to guide it?

Some answer has already been given to these questions in chapter 4, in the discussion of discovery processes. We saw there that search guided by only the most general heuristics of "interestingness" or novelty is a fully realizable activity. This kind of search, which provides the mechanism for scientific discovery, may also provide the most suitable model of the social design process.

It is generally recognized that in order to acquire new tastes in music, a good prescription is to hear more music; in painting, to look at paintings; in wine, to drink good wines. Exposure to new experiences is almost certain to change the criteria of choice, and most human beings deliberately seek out such experiences.

A paradoxical, but perhaps realistic, view of design goals is that their function is to motivate activity which in turn will generate new goals. For example, when about fifty years ago an extensive renewal program was begun in the city of Pittsburgh, a principal goal of the programe was to rebuild the center of the city, the so-called Golden Triangle. Architects have had much to say, favorable and unfavorable, about the esthetic qualities of the plans that were carried out. But such evaluations are largely beside the point. The main consequence of the initial step of redevelopment was to demonstrate the possibility of creating an attractive and functional central city on this site, a demonstration that was followed by many subsequent construction activities that have changed the whole face of the city and the attitudes of its inhabitants.

It is also beside the point to ask whether the later stages of the development were consistent with the initial one—whether the original designs were realized. Each step of implementation created a new situation; and the new situation provided a starting point for fresh design activity.

Making complex designs that are implemented over a long period of time and continually modified in the course of implementation has much in common with painting in oil. In oil painting every new spot of pigment laid on the canvas creates some kind of pattern that provides a continuing source of new ideas to the painter. The painting process is a process of cyclical interaction between painter and canvas in which current goals lead to new applications of paint, while the gradually changing pattern suggests new goals.

The Starting Point

The idea of final goals is inconsistent with our limited ability to foretell or determine the future. The real result of our actions is to establish initial conditions for the next succeeding stage of action. What we call "final" goals are in fact criteria for choosing the initial conditions that we will leave to our successors.

How do we want to leave the world for the next generation? What are good initial conditions for them? One desideratum would be a world offering as many alternatives as possible to future decision makers, avoiding irreversible commitments that they cannot undo. It is the aura of irreversibility hanging about so many of the decisions of nuclear energy deployment that makes these decisions so difficult.

A second desideratum is to leave the next generation of decision makers with a better body of knowledge and a greater capacity for experience. The aim here is to enable them not just to evaluate alternatives better but especially to experience the world in more and richer ways.

Becker and Stigler have argued that considerations of the sort I have been advancing can be accommodated without giving up the idea of fixed goals.[5] All that is required, they say, is that the utilities to be obtained from actions be defined in sufficiently abstract form. In their scheme the utility yielded by an hour's listening to music increases with one's capacity for musical enjoyment, and this capacity is a kind of capital that can be increased by a prior investment in musical listening. While I find their way of putting the matter a trifle humorless, perhaps it makes the idea of rational behavior without goals less mysterious. If we conceive human beings as having some kind of alterable capacity for enjoyment and appreciation of life, then surely it is a reasonable goal for social decision to invest in that capacity for future enjoyment.

Designing as Valued Activity

Closely related to the notion that new goals may emerge from creating designs is the idea that one goal of planning may be the design activity itself. The act of envisioning possibilities and elaborating them is itself a pleasurable and valuable experience. Just as realized plans may be a source of new experiences, so new prospects are opened up at each step in the process of design. Designing is a kind of mental window shopping. Purchases do not have to be made to get pleasure from it.

One of the charges sometimes laid against modern science and technology is that if we know *how* to do something, we cannot resist doing it. While one can think of counterexamples, the claim has some measure of truth. One can envisage a future, however, in which our main interest in both science and design will lie in what they teach us about the world and not in what they allow us to do to the world. Design like science is a tool for understanding as well as for acting.

Social Planning and Evolution

Social planning without fixed goals has much in common with the processes of biological evolution. Social planning, no less than evolution is myopic. Looking a short distance ahead, it tries to generate a future that is a little better (read "fitter") than the present. In so doing, it creates a new situation in which the process is then repeated. In the theory of evolution there are no theorems that extract a long-run direction of development from this myopic hill climbing. In fact evolutionary biologists are extremely wary of postulating such a direction or of introducing any notion of "progress." By definition the fit are those who survive and multiply.

Whether there is a long-run direction in evolution, and whether that direction is to be considered progress are of course two different questions. We might answer the former affirmatively but the latter negatively. Let me venture a speculation about the direction of social and biological evolution, which I will develop further in the next two chapters. My speculation is emphatically *not* a claim about progress.

From a reading of evolutionary history—whether biological or social—one might conjecture that there has been a long-run trend toward variety and complexity. There are more than a hundred kinds of atoms, thousands of kinds of inorganic molecules, hundreds of thousands of organic molecules, and millions of species of living organisms.

Mankind has elaborated several thousand distinct languages, and modern industrial societies count their specialized occupations in the tens of thousands.

I shall emphasize in the following chapters that forms can proliferate in this way because the more complex arise out of a combinatoric play upon the simpler. The larger and richer the collection of building blocks that is available for construction, the more elaborate are the structures that can be generated.

If there is such a trend toward variety, then evolution is not to be understood as a series of tournaments for the occupation of a fixed set of environmental niches, each tournament won by the organism that is fittest for that niche. Instead evolution brings about a proliferation of niches The environments to which most biological organisms adapt are formed mainly of other organisms, and the environments to which human beings adapt, mainly of other human beings. Each new bird or mammal provides a niche for one or more new kind of flea.

Vannevar Bush wrote of science as an "endless frontier." It can be endless, as can be the process of design and the evolution of human society, because there is no limit on diversity in the world. By combinatorics on a few primitive elements, unbounded variety can be created.

The Curriculum for Social Design

Our examination of the social planning process here suggests some extension of the curriculum for design that was proposed in the last chapter. Topic 7, the representation of design problems, must be expanded to incorporate the skills of constructing organizations as frameworks for problem representation, building representations around limiting factors, and representing non-numerical problems. Our discussion also suggests at least six new topics for the curriculum:

1. *Bounded rationality.* The meaning of rationality in situations where the complexity of the environment is immensely greater than the computational powers of the adaptive system.

2. *Data for planning.* Methods of forecasting, the use of prediction and feedback in control.

3. *Identifying the client.* Professional-client relations, society as the client, the client as player in a game.

4. *Organizations in social design.* Not only is social design carried out mainly by people working in organizations, but an important goal of the design is to fashion and change social organization in general and individual organizations in particular.

5. *Time and space horizons.* The discounting of time, defining progress, managing attention.

6. *Designing without final goals.* Designing for future flexibility, design activity as goal, designing an evolving system.

With the exception of control theory and game theory, which are of central importance to topics 2 and 3, the design tools relevant to these additional topics are in general less formal than those we described in the previous chapter. But whether we have the formal tools we need or not, the topics are too crucial to the social design process to permit them to be ignored or omitted from the curriculum.

Our age is one in which people are not reluctant to express their pessimism and anxieties. It is true that humanity is faced with many problems. It always has been but perhaps not always with such keen awareness of them as we have today. We might be more optimistic if we recognized that we do not have to solve all of these problems. Our essential task—a big enough one to be sure—is simply to keep open the options for the future or perhaps even to broaden them a bit by creating new variety and new niches. Our grandchildren cannot ask more of us than that we offer to them the same chance for adventure, for the pursuit of new and interesting designs, that we have had.

from: Simon, Herbert A. (1996), The Sciences of the Artifical, third edition, 139-141, 153-167,

NOTES

1 The authors of *The Federalist* were Madison, Hamilton, and Jay, but principally the first named. My edition is that edited by P. L. Ford (New York: Holt, 1898). Madison's notes are our chief source on the proceedings of the convention.
2 The notion of organizational survival and equilibrium depending on the balance of inducements and contributions is due to Chester I. Barnard, *The Function of the Executive* (Cambridge: Harvard University Press, 1938).
3 My views on some of these matters have been expounded at length in *H.A.* Simon, *Administrative Behavior,* 3rd ed., (New York, NY: The Free Press, 1976), H.A. Simon, V. A. Thompson and D. W. Smithburg, *Public Administration* (New Brunswick NJ: Transaction Publishers, 1991); and J.G. March and H. A. Simon *Organizations*, 2nd ed., (Cambridge, MA: Blackwell, 1993). On the nature of business organizanons, and especially the role of organizational identification in maintaining them, see chapter 2 of the present volume.
4 This section owes much to James G. March, who has thought deeply on these lines. See his "Bounded Rationality, Ambiguity, and the Engineering of Choice," *Bell Journal of Economics* 9(1978):587-608.
5 G. J. Stigler and G. S. Becker, "De Gustibus non est Disputandum," *American Economic Review*, 67(1977):76-90.

Charles Owen

4 BUILDING THE KNOWLEDGE BASE

Introduction

"Research" in design has a long but not very robust history. Individuals have published on the subject almost from the time design was recognized as something to be taught (engineering and architectural design theories have been in the literature since Roman times). Yet, despite exceptional efforts by some individuals, the degree of interest in research among the design disciplines has been quite uneven, ranging from more than a little in engineering design, to some in architectural and product design, to not very much in the fields of design most closely associated with the arts and crafts. In sum, in comparison to what is normally encountered in the sciences, humanities, and other scholarly disciplines, there has been precious little interest in what might be thought of as "classic" research.

But change is afoot. Events are propelling industries and countries into new economic relationships, and design is being recognized as a critical factor for business success. The result is new interest in the quality of design available, and—more fundamentally—interest in how design can be improved. As export strength commands more attention as an economic indicator, the improvement question becomes very important, its answer imperative.

For developed and developing countries alike, high-quality design is the most cost-effective resource available to improve trade balances. A few good designers using advanced design processes can have dramatic impact on the success of products and services. The obvious inference is that it behooves countries, industries, and companies to develop high-quality designers and equip them with high-quality design tools: theory, methods, and processes.

Thus, design research. And thus, among design educators, new interest in the nature of design research—especially as it may extend understanding beyond definitions of classic research used by the sciences and scholarly disciplines. In fields where the thrust of work is synthetic rather than analytic, this questioning is not naive. There is value in serious reflection on the most basic questions concerning research. What follows should be interpreted as such an exploration—an attempt to abstract from what we know in the hope of finding new models that may shed light on what we can do in design.

Figure 1
A Map of Disciplines.

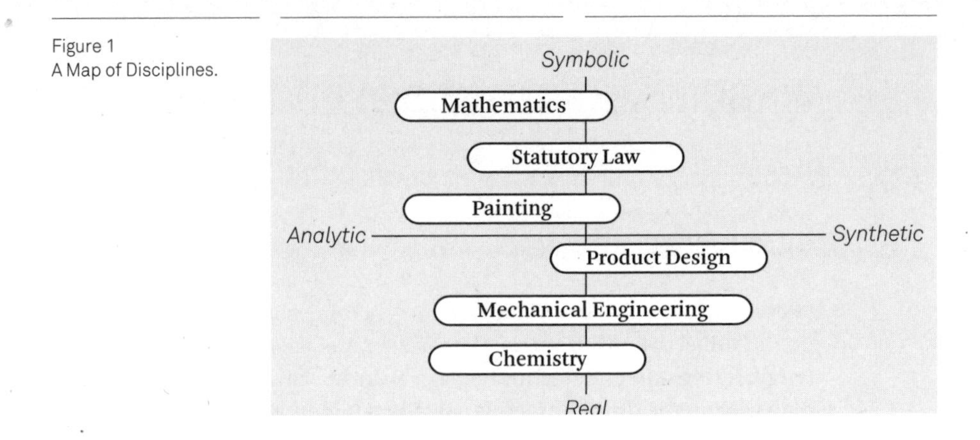

The Problem

Design, as a discipline, is still young (or, perhaps, is a slow learner). At any rate, it has not developed the internal structures and understanding that older disciplines have. Design is not science, and it is not art—or any other discipline. It has its own purposes, values, measures, and procedures. These become evident through comparisons, but they have not been extensively investigated, formalized, codified, or even thought much about in literature created for the field. In short, there is little to point to as a theoretical *knowledge base* for design. As a result, those who seek to work more rigorously look to scientific and scholarly models for guidance, and we find references to "design science" and examples of "design research" that would seem to fit more appropriately in other fields.

Yet, it is reasonable to think that there are areas of knowledge and ways of proceeding that are very special to design, and it seems sensible that there should be ways of building knowledge that are especially suited to the way design is studied and practiced. To approach these questions, it is probably best to abandon the term "research" for a time and, instead, look at how knowledge is used and accumulated—since building knowledge, after all, is the goal of research.

As a context for thinking about specialized knowledge acquisition and use, a *Map of Disciplines* reveals interesting differences among traditional fields of study and practice. Two axes define the map in Figure 1. Separating the map into left and right halves is an *Analytic/ Synthetic* axis. Disciplines positioned to the left of center are more concerned with "finding" or discovering; disciplines to the right are oriented toward "making" and inventing. A *Symbolic/Real* axis divides the map again into halves—vertically this time, according to the nature of the subjects of interest. Disciplines in the upper half of the map are more concerned with the abstract world and the institutions and communications that allow people to live and work together. Disciplines in the lower half work with the real world and the artifacts and systems that enable us to operate in the physical, not always friendly, environment.

A sample of disciplines illustrates how the map discriminates. In the upper half, mathematics, statutory law, and painting work with abstract, symbolic subjects; below, product design, mechanical engineering, and chemistry deal with real world phenomena. Mathematics, painting, and chemistry are primarily analytic in procedure; product design is almost entirely synthetic; and statutory law and mechanical engineering achieve something of a balance.

The positionings are, of course, subjective and relative, but they provide a means for gross comparisons on the basis of two very fundamental ideas about content and procedure.

The map is also a means for examining other relationships. Mechanical engineering seems nicely centered between the analytic and synthetic domains, but it is a discipline with subdisciplines. Engineering science, as one of these, would be located on the analytic side; engineering design would be more on the synthetic side. Hierarchical decompositions such as this afford opportunities for leveling or sharpening descriptions. Merging usually levels, moving the result of composition toward the center; decomposing sharpens, disseminating new elements into the quadrants.

Movements of disciplines over time can also be tracked. Through much of its history, painting was concerned with commissioned applications for clients. The trends of the last century moved it radically to the left, and it has become considerably more analytical and exploratory in intent and procedure.

No matter where they are on the map or how they move, merge, or diverge, all disciplines build knowledge bases. How they do this is important because it sheds light on the process and offers analogies for design. There is no single means, and the multiplicity strengthens the results.

Characterizing The Process

Knowledge is generated and accumulated through action. Doing something and judging the results is the general model. In Figure 2, the process is shown as a cycle in which knowledge is used to create works, and works are evaluated to build knowledge.

Knowledge using and knowledge building are not unstructured processes. They are controlled by *channels* that direct the procedures that are used to do and judge the work. These channels are the systems of conventions and rules under which the discipline operates. They embody the measures and values that have been empirically developed as "ways of knowing" as the discipline has matured. They may borrow from or emulate aspects of other disciplines' channels, but, in the end, they are special to the discipline and are products of its evolution.

The general model of Figure 2 can be extended to a model that fits the dual nature of actions suggested by the analytic/synthetic dimension of the Map of Disciplines. In Figure 3, this is done with an additional specialization of labels.

Figure 2
A general model for generating and accumulating knowledge.

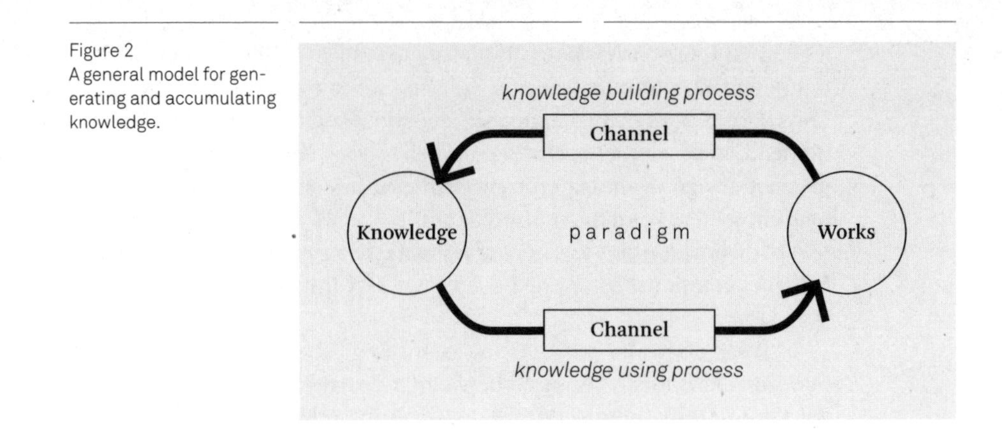

On the left side of the diagram, the realm of theory, the model is a paradigm for inquiry. Existing *knowledge,* under the direction of *theory,* is used to generate *proposals. Proposals* are tested with *measures* that verify or refute conclusions to build knowledge.

On the right side, the realm of practice, the model forms a paradigm for application. Here, *knowledge* is used through the application of *principles* to produce

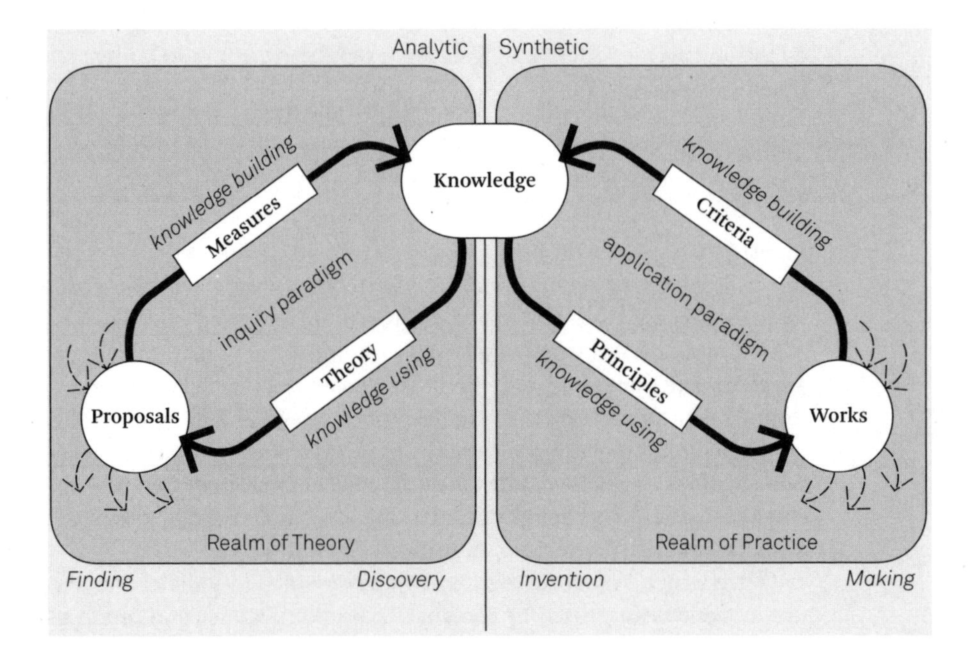

Figure 3. Using and accumulating knowledge in the two realms.

works. Works are judged for their worth as additions to the knowledge base using the *criteria* of the discipline.

Proposals and works also benefit from and contribute to ideas in other disciplines. A more complex diagram would show interdisciplinary relationships. Figure 3 suggests these as dashed arrows entering and leaving proposals and works.

Some Examples

To test the model, Figure 4 shows the sample disciplines of Figure 1 fitted with titles more expressive of their special characters. The darkness of the background suggests the skew of their primary activity to either the realm of theory or realm of practice—darker meaning more commitment.

It is hard to find a set of words that optimally fits a discipline—clearly fits it better than any other set of words—and differentiates it distinctly from other disciplines. Such nuance requires considerable variety and subtlety. Fortunately, both are available in English, and at least an attempt can be made. As an example, mathematics, for a paradigm of inquiry, *postulates propositions* using *axiomatic theory* and *proves* them with *reason* to build knowledge. In application, *models* are built with mathematical *principles* and *verified* with the *laws* of mathematics to add to applied knowledge. For better or worse, the other examples in Figure 4 similarly attempt to distinguish differences in procedure, objects of effort, and means of procedural control through choices of appropriate terms.

Selectively substituted words bring the generalized model into harmony with a specific discipline. Even though not perfect, they convey meanings well enough to convince. They also supply different viewpoints, a goal for extending our conception of knowledge-building processes in design.

Using And Building

In the acts of both doing and judging, questions are asked, answers obtained, and decisions made. How these are formed is the key to using knowledge successfully to build new knowledge.

Questions, answers, and decisions differ fundamentally in nature from discipline to discipline. They are framed from the value systems embedded in the disciplines. Table 1 suggests some of the these differences using the sample disciplines. Note that the differences are far deeper than issues of content. They grow directly from the basic values that create the knowledge structures of a discipline. As an interesting derivative of this comparison, it is possible to see through these differences the reason that design is not science or art, although it shares some of the characteristics of each.

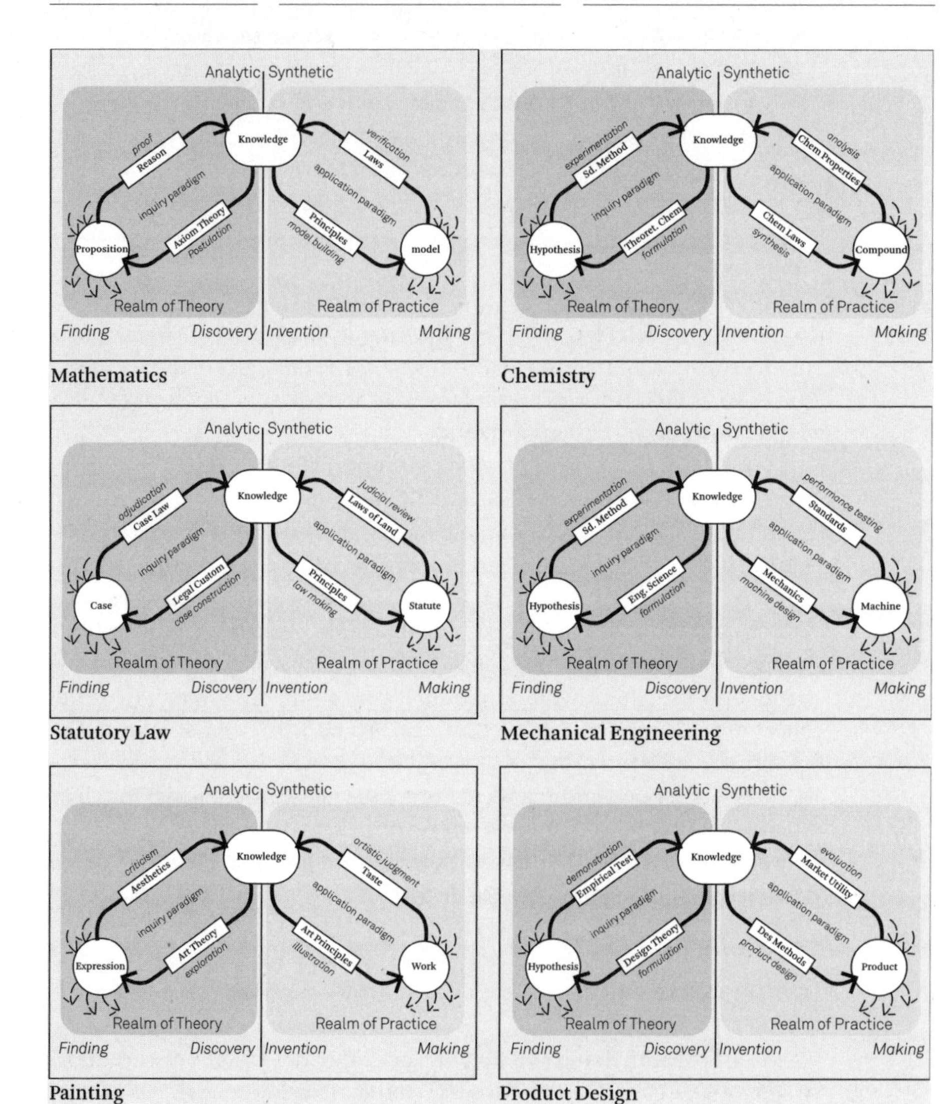

Figure 4
Sample disciplines with titles appropriate to their purposes

The forms of questions, answers, and decisions also differ *within* disciplines—between inquiry and application, and between doing and judging. These reflect the difference in purpose between inquiry and application and the difference in process between doing and judging.

For comparison purposes, the processes of using and building knowledge can be expressed as concatenations of question/answer and question/decision mediated through the channel appropriate to the process.

Table 1
Differences in Measures

Domain	Discipline	Measures	Source of Values
Science	Mathematics	true/false correct/incorrect complete/incomplete	reason logic
	Chemistry	true/false correct/incorrect right/wrong works/doesn't work	physical world
Technology	Mechanical Engineering	right/wrong better/worse works/doesn't work	physical world artificial world
Law	Statutory Law	just/unjust lawful/unlawful right/wrong	social contract
Arts	Painting	beautiful/ugly skillful/unskillful thought provoking/banal	culture
Design	Product Design	better/worse beautiful/ugly fits/doesn't fit works/ doesn't work	culture artificial world

Consider first the case of inquiry, the classic and most thoroughly discussed process. Here, the form of a question in knowledge using, or doing, is theoretical or methodological, seeking to find understanding of a phenomenon or process important to the discipline. An answer is formed as an evaluatable proposal. For the judging required for knowledge building, the form of question and decision is derived from the discipline's value system, setting a framework for judgment and measures to be used.

On the application side, doing involves questions and answers specific to a work or project that has been undertaken. Questions search for understanding of entities, relationships, and contextual elements within the project. Answers embody the understanding in ideas that draw on insights—solutions to problems. Judging, again, draws on the values of the discipline for the kinds of questions to ask and the criteria to make decisions. Questions thoughtfully constructed using these criteria exact decisions that determine a work's contribution to the knowledge base. The contribution, in this case, is the work or aspects of it that, through new syntheses, add to what is known about how the discipline's knowledge can be applied.

DESIGN RESEARCH

What light does all of this shed on the subject of design research? For a beginning, there are several general insights.

Some General Insights

First, research should not be thought of as being limited in form, in particular, to the classical forms of scholarly and scientific research. Those forms of research, as processes of knowledge using and building in the service of inquiry, are practiced by nearly all disciplines, but to greater or lesser extents. Knowledge using and building for the purposes of application is an equally productive process, adding to a discipline's knowledge base through the contribution of worked examples. A corollary lesson from this reflection is that balance may be useful.

Second, the processes of knowledge using and building are fundamentally the same for inquiry and application. The differences are more in the purpose of the activity. In both cases what is

[...] known is used to generate something new that will provide answers to questions inspired by a felt need. In the case of inquiry, the need is for deeper understanding of the subjects of a discipline; in the case of application, the need is for artifacts and institutions that employ the knowledge of the discipline more successfully.

Third, determinations of value must be understood to derive from the value system underlying a discipline. The kinds of questions framed by one discipline are not necessarily those of another. It is counterproductive, misleading, and a mistake, for example, to attempt to determine "rightness" or "truthfulness" within a discipline if these are not the relevant kinds of questions to ask.

Fourth, a position far to the left or right on the Map of Disciplines opens special opportunities for kinds of research appropriate to the other side. Disciplines skewed to the analytic side probably have unexplored opportunities for knowledge building through applications. Disciplines on the synthetic side should look to areas of inquiry—frequently the tools of the discipline (theory, methods, process) are worthy subjects for research.

Fifth, within the processes of framing questions and constructing answers or decisions lies the heart of good research and, ultimately, the basis for its quality. Questions sharply honed against the context of a discipline's value system require answers similarly crafted and decisions equally well constructed. Creativity, whether discovery or invention, is inspired by good questions.

Recommendations for Design

The design disciplines are on the synthetic side of the Map of Disciplines, far enough to the right that their claims to accomplishments in matters of inquiry are not extensive. This suggests a movement correction to the left for balance. Several other recommendations can be made; those following are primarily for design education, but a discipline includes practitioners, educators, researchers, and other associates with specialized responsibilities, so there are implications for many, including those responsible for collecting and disseminating design knowledge.

- Distinguish between research and professional advanced education. Graduate studies should be formalized to recognize the difference between studies to achieve mastery of the latest and most sophisticated design theory, methods, and process (application), and studies to create new design theory, methods, and process (inquiry). Degree titles can recognize these distinctions.

- Institute more structured programs of advanced study. Design has reached a level of maturity at which graduate courses can be taught with real information content. The master/apprentice model of an advanced degree course requiring only a longer, more thorough project is no longer adequate. Masters and doctoral programs with taught-course components are feasible and necessary.

- Define areas of design inquiry and application for which research is desired and establish funded centers and programs to accomplish the research. Design research has major potential value in a number of content areas—transportation, health care, information access, learning, work, urban systems, and design processes—to name just a few.

- Differentiate areas of design specialty and concentrate resources. Schools with specialized research programs can assemble equipment, financial and human resources synergistically to do better work than can be done with the same resources spread among many.

- Seek out faculty with research experience from disciplines related to design. To prime the pump, faculty members from other fields who are sympathetic to the goals of design can bring general research attitudes, procedures and rigor to the discipline. A few such interdisciplinary members will not dilute a design program, and their fresh ideas may well lead to useful evolutions in design research.

- Initiate studies of the philosophy of design. Just as studies of the philosophy of science, history, religion, etc. seek to understand the underpinning values, structures, and processes within these systems of knowledge building and using, there need to be studies of the nature of design. The design disciplines need thoughtful study of how design proposals and works are produced and evaluated. Measures and criteria as well as procedures for use and judgment should have the same attention given to scientific method. What is the analogous *design method?*

- Extend the means for communicating design knowledge. Most analytically oriented disciplines have extensive infrastructures of conferences, symposia, journals, text book publishing, and other communication systems that attract, collect, and distribute developing knowledge. These also act as recognition systems and create incentives for young faculty members to produce work of value.

- Inculcate knowledge-using more effectively in the question-asking phases of design applications. Design projects that have better thought-out beginnings will have better thought-out endings and, therefore, will be better candidates for building the experiential knowledge base.

Conclusions

Stepping away from the term *research* allows it to be seen more clearly. Asking instead how knowledge is built, widens the focus. With a broader view, it is possible to see how activities, seemingly opposed, actually work together to support the growth of knowledge.

A knowledge-using/knowledge-building model resolves the "who does research?" debate. Knowledge building is done in different ways, all of which contribute. In recognition of this, the Institute of Design has tailored its graduate programs to research and professional degrees (inquiry and application dimensions) paired with tracks for design planning and human-centered design (symbolic and real dimensions). Knowledge using and building are fundamental to the tracks in both programs.

The interest now being shown in design research is timely. Whether its inspiration is defensive (justifying educational budgets), competitive (contributing to an educational or industrial advantage), or simply idealistic (bringing the discipline to maturity), the impact will be the same. The health of our discipline will be well served by this needed attention to its foundations.

from: Design Studies 19, No. 1 (January 1998), 9-20, Copyright Elsevier

Christopher Frayling

5 RESEARCH IN ART AND DESIGN

Where artists, craftspeople, and designers are concerned, the word 'research'—the r word—sometimes seems to describe an activity which is a long way away from their respective practices. The spoken emphasis tends to be put on the first syllable the *re*—as if research always involves going over old territory, while art, craft, and design are of course concerned with the new. The word has traditionally been associated with; obscure corners of specialised libraries, where solitary scholars live; white-coated people in laboratories, doing esoteric things with test-tubes; universities, rather than colleges; arms length, rather than engagement; artyfacts, rather than artefacts; words not deeds.

Recently an opposing tendency has emerged—largely as the pragmatic result of decisions about government funding of higher education—where the word has come to be associated with: what artists, craftspeople, and designers do all the time anyway; artefacts, rather than artyfacts; deeds not words.

Much of the debate—and attendant confusion—so far, has revolved around a series of stereotypes of what research *is*, what it *involves* and what it *delivers*. The debate has also led towards some very strange directions indeed—such as the question (asked in all seriousness) 'does an exhibition of paintings *count* as research or doesn't it?' This paper attempts to unpack some of the stereotypes, and redirect the debate away from some of its more obviously blind alleys.

There seem to be almost as many definitions of research flying around in higher education right now, as there are reasons for promoting them. So I thought I'd go back to base—the OED. The *Oxford English Dictionary* lists two basic definitions, one with a little r and one with a big R, and within these, many many subsidiary ones. Research with a little r—meaning 'the act of searching, closely or carefully, *for* or after a specified thing or person'—was first used of royal genealogy in 1577, then in one of the earliest detective stories William Godwin's *Caleb Williams* in 1794 (where it concerned dues and evidence), then by Charlotte Bronte in 1847 to describe the search for overnight accommodation. Subsidiary definitions include 'investigation, inquiry into things; also a quality of persons carrying out such investigation' and, in music and poetry, 'a kind of prelude, wherein the composer seems to *search* or look out for the strains and touches of harmony, which he is to use in the regular piece to be played afterwards'. So research with a little r has been used, in the last four hundred years, of art practice, of personal quests, and of clues and evidence which a detective must decode. The point, says the OED, is that the search involves care, and it involves looking for something which is defined in advance: a criminal, a bed for the night, a regular musical theme. It isn't about professionalism, or rules and guidelines, or laboratories. It is about searching.

Research with a big R—often used in partnership with the word 'development'—means, according to the OED, 'work directed towards the innovation, introduction, and improvement of products and processes'. And nearly all the listed usages, from 1900 onwards, are from the worlds of chemistry, architecture, physics, heavy industry, and the social sciences. Research as professional practice, which earns it the big R. And its usage developed with the professionalisation of research in the university sector and in the chemical industry. In 1900, the word 'research' as applied to the humanities—for example—would have meant:

- antiquarianism
- the study of constitutional documents
- a self-motivated activity funded by paid teaching or other occupation.

The concept of humanities research as discovering new perspectives, or new information, is actually a very recent formulation.

So the dictionary doesn't take us very far—except that it establishes that the word has traditionally been used of art (and, with a big R, of design), and that for an activity to count in either sense as research the subject or object of research must exist outside the person or persons doing the searching. And the person must be able to tell someone about it. The dictionary also shows that prior to the turn of the century the word research carried no specific scientific meaning—indeed it predated the division of knowledge into arts and sciences.

But we aren't, of course, talking about *definitions* here: we're talking about usage, and this is where the Humpty Dumpty principle comes in. In *Alice Through the Looking Glass,* Humpty Dumpty has strong views about how words come to mean what they do:

> 'There's glory for you', said Humpty Dumpty.
>
> 'I don't know what you mean by 'glory'', Alice said.

> Humpty Dumpty smiled contemptuously. 'Of course you don't—till I tell you. I mean 'there's a nice knockdown argument for you!"
>
> 'But 'glory' doesn't mean 'a nice knock-down argument" Alice objected.
>
> 'When I use a word' Humpty Dumpty said, in rather a scornful tone, 'it means just what I choose it to mean—neither more nor less'.
>
> 'The question is' said Alice, 'whether you can make words mean so many different things'.
>
> The question is', said Humpty Dumpty. 'which is to be master—that's all'.

Which is to be master? Or, to put it another way, where does the legitimation come from? From a peer group, or an institution, or a funding structure, or an invisible College, or a section of society at large? Is this a political question, with a small p: about degrees and validations and academic status, the colour of peoples' gowns or, more interestingly, a conceptual one, about the very bases of what we all do in art, craft, and design?

Assuming that a little more than politics and money-raising are going on, I'd like to look at some of the widely shared assumptions which surround this debate—and, in unpacking them, at the ways in which its terms can perhaps be adjusted to be of more practical use.

And I'd like to start with Picasso's painting *Les Demoiselles d'Avignon.*

> 'In my opinion' said Picasso, 'to search means nothing in painting. To find is the thing. Nobody is interested in following a man who, with his eyes fixed on the ground, spends his life looking tor the pocket-book that fortune should put in his path . . .'
>
> Among the several sins that I have been accused of committing, none is more false than the one that I have, as the principal objective in my work, the spirit of research. When I paint, my object is to show what I have found and not what I am looking for. In art intentions are not sufficient and, as we say in Spanish, love must be proved by facts and not by reasons . . .
>
> 'The idea of research has often made painting go astray, and made the artist lose himself in mental lucubrations. Perhaps this

> has been the principal fault of modern art. The spirit of research has poisoned those who have not fully understood all the positive and conclusive elements in modern art and has made them attempt to paint the invisible and, therefore, the unpaintable.'[1]

In this rare interview of 1923 (one of only a few he ever gave for publication), Picasso is in part describing the *reference materials* he had used when preparing *Les Demoiselles d'Avignon* of 1906-7. Visual memories of the red-light district of Barcelona, some ancient Iberian sculptures he'd seen in the Louvre, Cezanne's Mont-Sainte-Victoire, a recent Matisse. But, he says, such reference materials should not *be* confused with research and in any case, the point of the exercise is single-mindedly to produce a finished painting. Only the art historians—a breed of which he was very suspicious—would think otherwise, after the fact. Yes, he had the spirit of research in him. But that was *not* his objective. Research to the painter, he said, equals visual intention. He's a maker not a researcher—and he doesn't even feel comfortable verbalising about his work.

The work may be ambiguous—even in 1923, there were squabbles about what it could possibly mean—but the artist isn't in the business of unambiguous communication. To adapt Herbert Read's famous distinction about art education, this is research *for* art, rather than either research *into* art or research *through* art, if indeed it is research at all. I will be elaborating on these distinctions a little later on.

But there's an even more dramatic moment, which neatly illustrates several popular assumptions about the artist's relation to his or her own practice—about how art happens. It is from a 1956 Hollywood film called *Lust for Life,* and it involves Kirk Douglas, sporting an orange beard, batting off some crows above a wheat-field in the South of France.

In this sequence, Vincent van Gogh played by Kirk Douglas, is impetuous, antirational, inward looking—and convinced that he is on an impossible quest to express what is on his mind, or in his mind's eye. He's white, male and quite barmy. He can't talk about his art—'it's impossible' is the best he can do—and he works very fast—just look at the way he paints all those crows, and creates his stormy sky of turbulent dark blue. The resulting picture becomes yet more evidence of his mental disturbance, which itself is evidence of something called his 'artistic temperament': *Crows over a Wheatfield* (in Auvers) is, according to the film, completed at great speed just moments before he tries to shoot himself—in the summer of 1890; actually, this *wasn't* his last canvas, but pop history has always preferred to believe that it was. The artist, by definition, is someone who works in an *expressive* idiom, rather than a *cognitive* one, and for whom the great project is an extension of personal development: autobiography rather than understanding. The movie stereotype of the artist is almost invariably like this—from Michelangelo, played by Charlton Heston (1965), or Caravaggio played by Nigel Terry in Derek Jarman's version (1986), to Ken Russell's biographies of more modern artists. And it is, I believe, shared by many outside our world. The question is, which is to be master, that's all. As John A. Walker has written:

> 'the idea that art might be a construction . . . rather than an expression, or that it might be the consequence of a host of social factors, is alien to the ethos of Hollywood.'[2]

It is unthinkable, he adds, that there could ever be a popular movie about a non-expressive artist such as Piet Mondrian.

Moving on to the designer, up until relatively recently the popular stereotype was rather different. Instead of the expressive artist, we have the pipe-smoking boffin who rolls up his sleeves (always his, incidentally) and gets down to some good, honest hands-on experimentation. From Leslie Howard in *The First of the Few* (1942), to Michael Redgrave in *The Dam Busters* (1955). The designer-boffin's very best moment of donnish understatement came in *The Dam Busters,* when the man from the ministry says to Dr. Barnes Wallis (Michael Redgrave): 'Do you really think the authorities would lend you a Wellington bomber, for tests? What possible argument could I put forward to get you a Wellington?' To which the boffin replies, 'Well if you told them I designed it, d'you think that might help?' Cut to Barnes Wallace in the cockpit of a Wellington . . .

Doing is designing for these people—not systematic hypotheses, or structures of thought or orderly procedures; but potting-shed, hit-and-miss, sorry I blew the roof off but you know how it is darling, craft-work.

More recently, there's been a change in the popular image of the designer—reflected in assorted television advertisements of the late 1980s which show young designers at work or play. The designer is no longer a boffin but is now a solitary style warrior who knows his (still it is usually his) way around the inner city jungle, and who believes in an aesthetic of salvage, or junk.

The young designer has become an imagineer—an archeologist of images, and signs, and styles from within the urban wasteland. Not a creator of meaning so much as an intuitive searcher after the latest thing. Don't think twice, it's alright. I'm reminded of the designer who was overheard on a bus, saying, 'let's be philosophical about this, don't give it a second thought'.

Now, side-by-side with the image of the expressive artist, the boffin and the style-obsessed designer, we have the popular image of the research scientist, and how he or she works. My third public image. And it is almost at the opposite extreme to the mad artist and the trendy designer.

The research scientist is orderly, he—again, it tends to be *he*, in popular images—has conjectures and hypotheses and he sets about proving or disproving them according to a set of orderly procedures. His subject exists outside himself, so he must submerge his subjectivity and personality in order to study it. He takes a problem, makes tentative conjectures regarding the answer to it and keeps revising the answer in the light of neat, well ordered experiments, which must be repeatable or replicable. He is what is known as a critical rationalist.

Interestingly, this stereotype exists in pop depictions of true-life scientists, rather than fictional or fantasy ones. Studies of the image of the scientist in pop culture have shown how *fictional* scientists tend on the whole to be something else: lunatics, or alcoholics, or psychopaths, or obsessives of some description—I suppose Drs Frankenstein, Faustus, Jekyll and Strangelove are the prime examples—while real-life scientists in films tend on the whole to be impossibly saintly, incredibly generous, unbelievably humanitarian and very often martyrs to their staggeringly effective research as well—I suppose Edward G. Robinson as Dr. Ehrlich, the Nobel-prizewinner who discovered a cure for syphilis in *Dr Ehrich's Magic Bullet* (1941), or Greer Garson as Marie Curie in *Madame Curie* (1943) or Mickey Rooney as young Thomas Edison and Spencer Tracey as the grown-up one, are the classics. Edward G. goes out with an exhortation to the people in the 1 & 9s to 'rid men's hearts of the diseases of hatred and greed', while Madame Curie at the very moment of her scientific discovery turns to husband Walter Pigeon and says 'Pierre-do you mind? You look first' ('He smiles understandingly' says the script, and 'touches her arm'). But on the whole, the psychopaths have won the day The earliest animated story films, made by Georges Méliès in the first decade of this century, featured explorers and scientists as manic, top-hatted music-hall turns, belonging to something called the Institute of Incoherent Geography. Since then, it's been estimated that mad scientists or their creators have been the villains of 31% of all horror or fantasy movies worldwide, that scientific or psychiatric research has produced 40% of the *threats* in all horror and fantasy movies—and—by contrast—that scientists have only been the heroes of 11% of horror movies.

So it is saints and sinners.

The saints have a self-evidently 'scientific' way of thinking, they tend to say 'Eureka!', and their successes instantly persuade the scientific community around them of the wisdom of their ways. It all seems so simple. And yet, of course, critical rationalism, which relies on making everything explicit, by revealing the methods of one's logic and justifying one's conclusions, and which has at the heart of its enterprise a belief in clarity, has been under considerable theoretical attack in the last 10-15 years. Sociologists such as Harry Collins, in his book *Changing Order* and philosophers such as Paul Feyerabend, have stressed that in science—as in everything else—there may well be conjectures but many of them are unconscious and they tend to be changed or modified without any explicit discussion, and they tend to involve a significant measure of subjectivity. In other words, the Edward G. Robinson version of research doesn't much resemble what science looks like in the laboratory, or what it feels like to those who are doing it. *Changing Order,* according to Harry Collins, involves irrationality, craftsman's knowledge, negotiating reality rather than hypothesising about it, above all tacit knowledge rather than propositional knowledge (and when there *is* propositional knowledge, a fair amount of tacit knowledge is in *there,* too). In the history and philosophy of science, historians such as David Gooding—who studies the methods of Michael Faraday—are now stressing the links between experimental scientists and creative artists (through the joint uses of imagination, intuition and craft practice), especially in the nineteenth century. Where the artist has difficulty persuading people of the connection of art with research, the scientist (whose research expertise has until recently been taken for granted) has exactly the same problem with creativity—which is generally seen as the prerogative of the artist rather than the scientist. This is partly why the *process* of discovering has been virtually ignored until recently, and why the activity of fine art is of increasing interest to historians of science Look at *The Double Helix:* it could almost be an artist's autobiography.

If the stereotype of the *scientist* as researcher needs some adjusting—to make it seem *closer* to art and design (though by no means identical with it)—the popular image of the *fine artist* needs a lot of work as well. For, in the history of art since the Renaissance, there are of course countless examples of artists who have explored their materials for what they are, and not simply as 'raw materials'. Who have worked in a cognitive rather than an expressive idiom. George Stubbs's researches on animal anatomy—involving portfolios of drawings of dissections, which were also used by scientists—made possible George Stubbs's animal paintings and they have lived on in parallel with the pictures. John Constable's researches into cloud formation—his many cloud drawings and paintings—made possible John Constable's landscape paintings. This is not to suggest that Stubbs and Constable were, respectively, vet and weatherman, but that they operated quite consciously—in a cognitive idiom, researching subjects which existed outside themselves and their own personalities. In this century, one could cite artists who explore the doors of perception such as op artists—or computer artists—or artists as semiologists—as their heirs in this sense. Research *for* art and sometimes research *through art,* to re-use the distinction. One problem is, that the classic examples of this—Leonardo, Stubbs, Constable—date from a long time ago. Their drawings would be unlikely to be at the cutting edge of such research today, in the era of electron-microscopes and other ways of enhancing the image

As Tom Jones has written:

> 'While Leonardo da Vinci's drawings pioneered anatomical research, any work an artist does now in this vein can *only* be reference material, the study of anatomy having progressed far beyond what can be observed by the unaided eye. Additionally, the medical skills now required are so specialised that they are unlikely to be possessed by any artist. Indeed given current scientific understanding, it is difficult to conceive that much research into subject-matter (in the sense in which it has been defined relative to Stubbs, Constable and Leonardo da Vincil is possible nowadays.'[3]

It is much more likely to be a matter of referencing the subject or illustrating it in ways that photography cannot achieve.

Nevertheless, the examples show:

- that artists have worked just as often in the cognitive idiom as the expressive

- that some art counts as research—anyone's definition

- that some art doesn't.

It is a relief to discover that there's no question of giving every single painter since the Renaissance an honorary PhD, *in absentia*. Whatever definition we end up with, it can never in my view—in principle or in practice—fit all fine art activities. Why should it? If Picasso had wanted a doctorate of philosophy. I'm sure he would have registered for one. Instead he is said to have turned down honorary degrees all over the western world. There must be an institutional, or pedagogical, or academic, or technical, or some reason for wanting to do research. Not just status, promotion and fund-raising.

To illustrate this, here's a famous quotation from John Constable, to set against the Picasso statement I quoted earlier. The quotation is from a lecture to the Royal Institution in May 1836:

> 'I am here on behalf of my own profession, and I trust it is with no intrusive spirit that I now stand before you, but I am anxious that the world should be inclined to look to painters for information on painting. I hope to show that ours is a regularly taught profession, that it is scientific as well as poetic; . . . and to show by tracing the connecting links in the history of landscape painting that no great painter was ever self-taught. . . Painting is a science, and should be pursued as an inquiry into the laws of nature. Why, then, may not landscape be considered as a branch of natural philosophy, of which pictures are but experiments?'[4]

If the stereotype of the artist is fairly wide of the mark, the recent image of the young designer—descended from the image of the art student in general, which was invented as recently as the 1950s—also needs substantial readjustment. Not just in the light of what we know about design research, the design methods movement, basic design, and the whole range of attitudes towards the use of reference materials and procedures and mental attitudes—but, again, in the light of history. In a sense, the concept of design as research—either *applied research*, where the resulting knowledge is used for a particular application, or *action research*, where the action is calculated to generate and validate new knowledge or understanding, or even (but very rarely) *fundamental research*—is so well established that it doesn't need elaborating here. But popular assumptions about design—and indeed some of the self-images of designers—do live on. And what's less well known, is the fact that if you examine the origins of art and design teaching in Britain, you'll probably see that 'research' as a problem area, or as something which exists *outside* studio design, is, again, a relatively recent phenomenon. Let's take your average design student at the government school of design in London from the late 1840s to the 1860s. Already, art and design had been separated from the mainstream university sector—in 1836, they were poised to go in, but the Mechanics Institute-style mode! was adopted instead—but the curriculum was based to a large extent on formal rather than tacit knowledge, and on design as a kind of language. You learned the grammar—from books by Owen Jones, or papers by Gottfried Semper—and, if you were very lucky, you then learned the usage as well. But in studying the grammar—with reference to other grammars, such as those of botany and sometimes physics and mechanics—you were given access to the very latest research into the design process. It wasn't *doing* versus *thinking*. It was practice as an amalgam of the two, with, if anything, the emphasis on the thinking. Time enough to implement the thoughts after leaving College, it was thought.

To recapitulate:
The popular image of the fine artist as expressive lunatic does not allow sufficiently for the cognitive tradition in art—a tradition which has in fact been called 'research'. Nor does it allow for the fact that art happens in a social, technical, and cultural world.

The popular image of the designer as style warrior—superficial, trendy, obsessed with surfaces and signs—does not allow sufficiently for the research and methods tradition in design, or indeed for the tacit use of those methods by designers—to say nothing of applied semiotics. I once asked an eminent advertiser, while I was making my television series *The Art of Persuasion,* for Channel 4, about his line on the science of semiotics. 'Oh', he said 'that. That's what I do for a living!'

Equally, the popular image of the art and design student ignores those important moments in our history when research—in anyone's definition—was a central part of the curriculum.

By the same token, the popular image of the scientist—as critical rationalist, engaged in fundamental research and shouting things like 'Eureka' or 'it's a crazy idea but it just might work'—the image against which a lot of research tends still to be judged, is equally wide of the mark. Doing science—as opposed to post-rationalising about science—just doesn't seem to be like that, if recent researches into the philosophy and sociology of science are any guide. Doing science is much more like doing design.

Implicit in much of what I've been saying, is a criticism of yet another stereotype—that of 'the practitioner'. As if action which follows reflection, or reflection which follows action, can be put in a box exclusively marked 'practice'. Research is a practice, writing is a practice, doing science is a practice, doing design is a practice, making art is a practice. The brain controls the hand which informs the brain. To separate art and design from all other practices, and to argue that they alone are in a different world, is not only conceptually strange, it may well be artecidal (to use Stuart Macdonald's word), yes, art and design have been taught separately from the mainstream, ever since 1837. But that is an institutional accident, not a conceptual statement.

So, where does all this lead? Apart from to the important thought that 'research' is a much less diffuse, much more convergent activity than the terms of the recent debate would suggest. And that 'research' has been, can be and will continue to be an important—perhaps the most important—nourishment for the practice and teaching of art, craft and design.

There *is* a lot of common ground. There is also a lot of private territory. I'd like to finish with the three categories (derived from Herbert Read) with which I began, to make some practical suggestions as to the kinds of research which might suit, indeed grew out of, what we actually do;

- Research into art and design
- Research through art and design
- Research for art and design

Research *into* art and design is the most straightforward, and, according to the Allison index of research in art and design—as well as CNAA lists of the 1980s and early 1990s plus my own experience at the Royal College of Art—by far the most common:

- Historical Research
- Aesthetic or Perceptual Research
- Research into a variety of theoretical perspectives on art and design—social, economic, political, ethical, cultural, iconographic, technical, material, structural, . . . whatever.

That is research *into* art and design. At the College, it involves PhD theses or MPhil dissertations. And it is straightforward, because there are countless models—and archives—from which to derive its rules and procedures.

Research *through* art and design which accounts for the next largest category (though a small one) in the Allison index, the CNAA documents, and my own experience of degrees by studio project at the College, is less straightforward, but still identifiable and visible.

- materials research—such as the titanium sputtering or colorization of metals projects successfully completed in the metalwork and jewellery departments at the College and Camberwell, in association with Imperial College of Science & Technology (partnerships are very useful, in this area of research).

- development work—for example, customising a piece of technology to do something no-one had considered before, and communicating the results. A recent example: the Canon colour photocopier at the Royal College of Art, successfully used by some postgraduate illustration students, who have both exhibited and written up the results

- action research—where a research diary tells, in a step-by-step way, of a practical experiment in the studios, and the resulting report aims to contextualise it. Both the diary and the report are there to *communicate the results,* which is what separates *research* from the gathering of reference materials. Kenneth Agnew has recently and wisely written that research through the design of products has been

  'hindered by the lack of any fundamental documentation of the design process which produced them. Too often, at best, the only evidence is the object itself, and even that evidence is surprisingly ephemeral. Where a good sample of the original product can still be found, it often proves to be enigmatic'.[5]

These types of research resemble Herbert Read's 'teaching through art'—so long as we're clear about what is being achieved and communicated *through* the activities of art, craft or design. At the Royal College of Art, this kind of research, sometimes known as the degree by project—with a specific project declared in advance of registration—involves for the MPhil studio work and a research report, and for the PhD studio work plus a more extensive and substantial research report.

The thorny one is Research *for* art and design, research with a small 'r' in the dictionary—what Picasso considered was the gathering of reference materials rather than research proper. Research where the end product is an artefact—where the thinking is, so to speak, *embodied in the artefact,* where the goal is not primarily communicable knowledge in the sense of verbal communication, but in the sense of visual or iconic or imagistic communication. I've mentioned the cognitive tradition in fine art, and that seems to me to be a tradition out of which much future research could grow: a tradition which stands outside the artefact at the same time as standing within it. Where the expressive tradition is concerned, one interesting question is why people want to call it research with a big 'r' at all—What's the motivation? True, research has become a political or resource issue, as much as an academic one. And, as a slight digression, it always amuses me to see the word 'academic' used as a pejorative—by people who themselves earn their livings within the academy. Research has become a status issue, as much as a conceptual or even practical one.

And that—I must confess—worries me. There may well be opportunites for research within the expressive tradition, but they need dispassionate research rather than heated discussion about status, class and reverse snobbery.

At the College, we give *Higher Doctorates* or *Honorary Doctorates* to individuals with a distinguished body of exhibited and published work—but we do not at present offer research degrees entirely for work where the art is said to 'speak for itself'. Rightly or wrongly, we tend to feel the goal here is the art rather than the knowledge and understanding. The Picasso philosophy. And we feel that we don't want to be in a position where the entire history of art is eligible for a postgraduate research degree. There must be some differentiation.

- Research into art and design
- Research through art and design
- Research for art and design

The novelist E.M. Forster's aunt once said to Forster:

> 'How can I tell that I think till I see what I say?'

That seems to me to be very like the first category If we modify this to

> 'How can I tell what I think till I see what I make and do?',

then we've covered the second category as well But if we modify it further to

> 'How can I tell what I am till I see what I make and do?'

it seems to me we have a fascinating dilemma on our hands. As much about autobiography and personal development as communicable knowledge. I can only add, that research for art, craft and design needs a great deal of further research. Once we get used to the idea that we don't need to be scared of 'research'—or in some stange way protected from it—the debate can really begin.

from: *Royal College of Art Research Papers*, Volume 1, Number 1, (1993/4), 1-5,

REFERENCES

1 Pablo Picasso: an interview (reprinted from *The Arts*. New York, May 1923, in *Artists on Art,* ed. Robert Goldwater & Marco Treves. London: John Murray, 1985, 416-7)
2 John A Walker: Art *S Artists on Screen* (Manchester University Press, 1993, 46).
3 Tom Jones: *Research in the Visual Fine Arts* (Leonardo, 13,1980, 89-93).
4 John Constable: Lecture *notes.* May 26 & June 16 1836 (in Artists on Art, op.cit, 270-273).
5 Kenneth Agnew: 'The Spitfire: Legend or History? An argument for a new research culture in design' (*Journal of Design History* 6, 2, 1993: 121-130).

Bruce Archer

6 THE NATURE OF RESEARCH

There is more than one way of defining research, and there are several traditions as to how research should be carried out. I will try to describe the nature of research in terms that would be common to most of them.

First, very briefly, I will define research in its most general sense. Then I will talk about research, according to the Science tradition. Third, I will say something about research in the Humanities tradition. Finally, I will talk about research through Practitioner Action.

Research in General

Research is systematic enquiry whose goal is communicable knowledge:

- systematic because it is pursued according to 'some plan';
- an enquiry because it is seeks to find answers to questions;
- goal-directed because the objects of the enquiry are posed by the task description;
- knowledge-directed because the findings of the enquiry must go beyond providing mere information; and
- communicable because the findings must be intelligible to, and located within some framework of understanding for, an appropriate audience.

Research in the Science Tradition

In the Science tradition, several distinctive categories of research activity are recognised, the categories being distinguished by their intentionality. Widely accepted categories are:

(i) Fundamental Research: Systematic enquiry directed towards the acquisition of new knowledge, without any particular useful application in view.

(ii) Strategic Research: Systematic enquiry calculated to fill gaps in Fundamental Research and/or to narrow the gap between Fundamental Research and possible useful applications.

(iii) Applied Research: Systematic enquiry directed towards the acquisition, conversion, or extension of knowledge for use in particular applications.

(iv) Action Research: Systematic investigation through practical action calculated to devise or test new information, ideas, forms, or procedures and to produce communicable knowledge.

(v) Option Research: Systematic enquiry directed towards the acquisition of information calculated to provide grounds for decision or action.

The greatest volume of research in the Science tradition is categoriseable as Applied Research.

Applied Research may or may not result in inventions and discoveries of any significance, and its findings may or may not be widely generalisable. Indeed, Applied Research can often produce indeterminate or even completely empty results. Empty results are not necessarily valueless, In fact it may be very useful indeed to a manufacturer or a government department to know that a particular line of proposed development would be fruitless for themselves and/or for their competitors.

Action Research is often conducted by practitioners of one or other of the useful arts, such as medicine, teaching, or business, or, indeed, of any of the other disciplines embraced by design education, rather than by professional researchers. Here, I shall be using the term 'practitioner' to cover all or any of these. I shall be having a lot more to say about practitioner research and Action Research later on.

Option Research is more limited in scope, being pursued only to the point where sufficient information has been produced to enable the manager or policymaker who commissioned it to take a decision on the given issue. Most Option Research is highly situation-specific, that is, valid only in the circumstances of the situation enquired into.

Conducted properly it is nevertheless a systematic enquiry whose goal is knowledge. Much market and business research falls into this category. Very little Option Research is ever published, however. Consequently, it is regarded with the greatest suspicion by orthodox scientists. It is widely regarded as ineligible for the title 'research', and is rarely recognised as suitable for registration for a research degree.

Science is concerned with explanation. What can be observed? What events can be recorded? How does this, that, or the other event proceed? What is the cause of this or that? The scientific ideal, not always achieved, is to produce explanations that have enduring validity. Most particularly, the scientific ideal is to produce explanations that remain valid when tested in wider and wider fields of application, and which therefore offer some powers of prediction.

Science has many branches, from anthropology to astrophysics. However, Science is not defined by its subject matter. Science is defined by its intellectual approach. The range of subjects that have been addressed by Science, and which have ceased to be preserve of myth, fable, and metaphysics, has continued to expand throughout the history of mankind's endeavours. Scientists have the right to turn their minds to anything, as long as they do it scientifically.

But what do we mean by 'doing it scientifically'? The traditional Western perception of a correct scientific approach, still held by many to the present day, is based on ground rules that were first systematically set out by Francis Bacon in 1620. The whole process is characterised as being empirical (that is, based upon evidence obtained in the real world), objective (that is, free from the influence of value judgements on the part of the observer), and inductive (that is, moving from the observation of specific instances to the—formulation of general laws). Intellectual processes of any sort that fail the tests of empiricism, objectivity, and inductive reasoning are dismissed as unscientific and unreliable. So goes the Baconian paradigm.

However, amongst philosophers of science, and amongst scientists working at the more advanced levels of scientific endeavour, the last three decades have seen a complete overturning of the Baconian paradigm for the conduct of scientific enquiry. Chief amongst the revolutionaries has been Karl Popper, mathematician and philosopher, who is living today. His not very numerous, but highly influential, works on the philosophy of science have been published in English at intervals since 1959 (his earliest works, from 1934, were written in German), They have had a profound effect that reaches far beyond the bounds of conventional science. In them, he has rejected the whole traditional Baconian view of scientific method, and replaced it with another. His argument begins by pointing out the logical asymmetry between verification and falsification. His own well-known example runs like this:

> *No number of observations of white Swans allows us logically to derive the universal statement: 'All swans are white'. Searching for, and finding, more and more white swans does not prove the universality of the white swan theory. However, one single observation of a black swan allows us logically to derive the statement 'not all swans are white'. In this important logical sense, generalisations, although never verifiable, are nevertheless falsifiable. The pursuit*

of verification can go on forever, but falsification is instantaneous. This means that whilst most scientific theories are unprovable, they are still testable, indeed, the only reliable way to test a scientific proposition is to formulate it in as unambiguous a way as possible, 'and then to conduct systematic attempts to refute it.

It is this fast statement which lies at the heart of the Popperian revolution. Falsification of theory, not verification, should be the aim of scientific enquiry.

Also central to the Popperian view is the acknowledgement that new scientific propositions may properly be, and mostly are, the result of inspired guesswork rather than the product of inductive reasoning. By the same token, empiricism is more important in the stages when a theory is under test than in the stages when it is being formulated. The title of Popper's most influential book, first published in 1963, but revised several times since, is *Conjectures and Refutations*. This title encapsulates the Popperian view of the correct scientific approach. Empiricism, objectivity and induction have their place, but this is after tie formulation of an explanatory conjecture, not before it.

So how can we summarise the philosophy behind the modern approach to scientific research? The post-Popperian approach demands' that the investigator:

- be liberal about the sources of conjecture and hypothesis at the commencement of research; and

- be sceptical in the handling of data and argument; research; and

- be astringent in testing findings and explanations on the completion of research.

Research in the Humanities Tradition

In most English speaking countries, and certainly in Britain, the term *The Humanities* and the term *The Arts* tend to be a little bit confused, and are used almost interchangeably. Either term can be used to refer to that large group of academic disciplines in which mankind is the central concern, in contrast with *The Sciences*, in which the physical world is the central concern. I tend to employ the expression *The Humanities* as the umbrella term, and divided the disciplines within it into two subgroups; *Metaphysics*, comprising theology, philosophy, epistemology, ethics, aesthetics, etc.; and *The Arts*, comprising language, literature, drama, history, architecture, art, music, etc.

With reference to The Arts, I want to make a distinction between:

- the *practice* of the Arts, such as creating new works of literature, drama, music, etc., or performing existing works of drama, music, etc.;
- *scholarship* in the Arts, such as knowing the content, authorship, history, and categorisation of works in the Arts; and
- *research* into, or for the purposes of, Arts activity.

This section of my talk is concerned with academic attainment in Arts disciplines, and particularly with the award of research degrees in the Arts.
The disciplines of the Arts are variously concerned with:

- expression in appropriate media;
- creative reflection on human experience;
- the qualitative interpretation of meaning in human expression;
- judgements of worth;
- the exploration of truth values in text;
- the categorisation of ideas, people, things and events;
- the tracing of, and commentary upon, the provenance of ideas, people, things and events.

Some, but not all, Arts activities are based on empirical evidence in the real world. Some, but not all, Arts activity cites exemplars in the real world or in previous writings in support of argument leading to a postulated conclusion. Nevertheless, virtually all Arts activity is essentially subjective in character.

Scholarship in the Arts makes an important distinction between primary sources of information, and secondary sources of information. Primary sources include: originals, or original records of, or contemporaneous commentary upon, ideas, things, events or persons. Secondary sources include other persons' commentaries upon primary material. Acknowledged scholars in the Arts are expected to have a comprehensive knowledge of the primary sources in their field. They are expected to have a clear vision of the provenance of the important ideas, things and events. They are expected to offer critical appraisal of the more significant secondary source material. It is not necessary, however, for a scholar to have produced new primary material in order to be recognised as an authority in a given field,

In view of the subjective nature of Arts activity, any witness of a particular Arts work needs to know from which standpoint the author produced it. A popular challenge put to an author of either primary or secondary material by a witness on first confrontation with a work is: "What is your theoretical position?" This is an important question. The author's ideology and framework of values will have coloured his or her view of events, and will be embodied in his or her expression of them. Unless the witness shares the author's position, or at least recognises what that position is, he or she will not be able fully to understand the work or to judge it.

The use of the term 'ideology' in this context needs a bit of explanation. A formal definition of the term 'ideology' is:

> *An overarching system of explanation or interpretive scheme that serves to make the 'world' more intelligible to those who subscribe to it.*

Thus, the possession of an ideology enables a person of a community to make coherent sense of otherwise disconnected theories and experiences and values arising in different aspects of life. A political ideology can do this. So can a religious ideology. So can an aesthetic or a scientific ideology. Witnesses may or may not share a particular author's ideology in their hearts when confronted by an Arts work, but if they know what the author's position is, they can at least appreciate what the author was expressing. By the same token, it is the duty of a scholar in the Arts to make clear the standpoint from—which he or she may be offering opinion or discoursing upon the content, value and authorship of primary and secondary source material.

Master's degrees and Doctoral degrees, not being research degrees, may be awarded for the attainment of scholarship in the Arts.

The conduct of research in Humanities disciplines goes beyond scholarship. Scholarship is essentially comprehensive knowledge of a particular field in a particular discipline. Research in that discipline consists in finding new things to know, or in identifying new ways of knowing them, or in refuting previous commentary on existing material. In recent times, the once unbridgeable differences between Science research and Humanities research have moved closer together. Whilst Science still seeks ultimately to explain and Humanities still seek ultimately to evaluate, Science has become less reductionist in its attitudes and the Humanities more empirical. Moreover, their mutual use of databases and information technology has brought their methodologies closer together.

There was a time when in some Arts disciplines it was only necessary to uncover some hitherto unknown or unrecognised or unorganised material (say a complete set of the laundry lists of some dead poet) and to catalogue this material in order to earn a research degree. There are few areas where such a condition could apply today. Generally, research in the Humanities tradition advances by the conduct of logical argument. Propositions are validated or refuted by exemplification and citation.

It is in the nature of the Humanities disciplines that their judgements are made within a framework of values. There is no such thing as 'objective' Humanities research. That is why it is so important for the investigator to declare his or her 'theoretical position'. Nevertheless, some Humanities research strives to present findings generalisable within a given context. In such a case, it is up to the reader to determine whether or not the argument and the findings remain valid in that or a different context.

All the Humanities disciplines have well established rubrics for sound scholarly argument and good research practices. Common amongst them are requirements to distinguish between evidence produced by the research, evidence imported at second-hand, the judgements of others, and judgements by the investigator. Scholars are expected always to be alert to the pitfalls of circular argument, that is, where the author is seen to be saying that A is greater than B, while B is greater than C, and C is greater than A. Harder to spot, in a long and complex argument but equally condemned, are commutative arguments, that is, where A is said to be greater than B because B is less than A. Good practice in Humanities disciplines demands that all citations shall be checked at primary source level, and that all sources shall be acknowledged.

The principal purpose of pursuing an MPhil or MRes degree programme in the Arts, as in the Sciences, is to learn the methods of research appropriate to a given field of enquiry; to advance knowledge in a given discipline; and to qualify for admission to a PhD degree programme. The distinguishing features of an MPhil or MRes programme are:

- the critical appraisal by the candidate of prior research; and
- close attention to the principles and practice of research methodology;
- and the conduct under supervision of a single major task of systematic investigation.

MPhil and MRes programmes have, or should have, a substantial 'taught' element.

Where a PhD degree programme is offered, its distinguishing features will be

(i) the critical appraisal by the candidate of prior research; and

(ii) close attention to the principles and practice of research methodology; and

(iii) the conduct of a single major systematic investigation; and

(iv) the delivery of a substantial contribution to knowledge.

It is a telling mark of the trend in academic thinking in recent years that good practice in academic research in the Arts can be expressed in much the same terms as the post-Popperian paradigm adopted in Science research:

- The enquiry must be calculated to expose new observations or new explanations; or it 'must seek' to falsify previous observations or explanations; and

- the theoretical position from which the investigation is approached must be made clear; and

- the chief questions to be addressed by the enquiry must be unambiguously expressed. It is not necessary to show that the problem posed, or the conjectures employed, arise from empirical data, nor that they have been arrived at by inductive reasoning; but

- primary sources must be cited for evidence employed, and where secondary sources are referred to, the provenance of the ideas—handled must be indicated; and

- any new data obtained must be recorded so as to be checkable by later observers; and all the procedures and argument employed must be transparent to later observers; and

- all initial and intermediate conjectures must be configured so as to lend themselves to attempts at refutation during the course of the study, and all ultimate findings and conjectures must lend themselves to attempts at refutation by subsequent investigators; and

- the record of the investigation and its findings must be published or otherwise exposed to critical appraisal by other investigators.

Research Through Practitioner Action

Some artists and designers, and some other creative practitioners, claim that what they ordinarily do is research. They argue that their art works or design products or other creative practitioner output constitutes new knowledge. Moreover, they claim that the act of publicly exhibiting, installing, manufacturing or distributing their works, constitutes publication. Therefore, they say, creative practitioner activity is synonymous with research activity. To what extent can such a claim be justified?

Undoubtedly, there is such—a thing as tacit knowledge, that is, a kind of knowing that is not separated, or separable, from the perception, judgement or skill which the knowledge informs. There will be some of that in all creative practitioner activity. Undoubtedly some knowledge can be transmitted by some works to other practitioners, and possibly to the population in general, when the work is 'published'. Undoubtedly, in some circumstances, a striking art work or a radically new product or other innovation can itself constitute new knowledge, tacit or otherwise, that can be highly significant leading to major changes in people's perceptions, circumstances and values. Clearly, too, a great deal of practitioner activity entails some research, of orthodox or unorthodox kinds, in support of the main thrust of the practitionership. It is not quite so certain, however, that the practitioner activity *itself* is quite the same as research activity, however much research it may have been supported by.

One has to ask, was the practitioner activity an enquiry whose goal was knowledge? Was it systematically conducted? Were the data explicit? Was the record of the conduct of the activity 'transparent', in the sense that a later investigator could uncover the same information, replicate the procedures adopted; rehearse the argument conducted, and come to the same conclusions? Were the data and the outcome validated in appropriate ways?

Most academic institutions can point to at least a few cases of practitioner activity where an effort has been made, successfully, to meet these criteria, so can a few studios, research institutes and professional consultancy offices. In these cases, practitioner activity can properly be equated with research, and should be recognised and rewarded accordingly. Where, however, any activity, whether it claims to be 'research' or not, fails to meet the criteria which define research activity as 'a systematic enquiry whose goal is communicable knowledge', it cannot properly be classed as research or equivalent to research.

However, identity between practitioner activity and research activity is not the only possible relationship. There are other relationships that are worth exploring. It can be useful to distinguish between research about practice; research for the purposes of practice; and research through practice.

Research about practice can be of many kinds. Art or design history, for example, and the analysis and criticism of the output of art or design activity, are orthodox Humanities subjects. Studies about art or design in relation to people and society fall within the Social Sciences. Studies about the materials and the processes, which are or could be used in, or specified by various kinds of art or design activity, fall within appropriate Science disciplines. Studies of the methodologies of art or design fall within the cross-cutting discipline of design research, which embodies several kinds of research. All studies *about* practice, if they are to be recognised as research studies, must employ the methods, and accord with the—principles, of the—class to which they happen to belong.

Similarly, research activity conducted for the purposes of contributing to other practitioner activities can also fall into any category of Science or Humanities, and must be practised according to the principles underlying that category. Where an investigation for the purposes of contributing to a practitioner activity is conducted according to the principles of its field, and is indeed a systematic enquiry whose goal is communicable knowledge, then the investigation can properly be called research. However, the fact that research 'for the purposes of' has underpinned a particular practitioner activity does not permit the practitioner activity itself to be described as research.

It is when research activity is carried out *through the medium of* practitioner activity that the case becomes interesting. There are circumstances where the best or only way to shed light on a proposition, a principle, a material, a process or a function is to attempt to construct something, or to enact something, calculated explore, embody or test it. Such circumstances occur frequently in explorations in, for example, agriculture, education, engineering, medicine and business. Such explorations are called Action Research, which I defined earlier as 'systematic enquiry conducted through the medium of practical action; calculated to devise or test new, or newly imported, information, ideas, forms or procedures and generate communicable knowledge'. The principles for the proper conduct of Action Research are well established in agricultural research, educational research, medical research, etc.; as well as in all or most of the disciplines of, for example, this Environment School.

All the normal rules governing research practice apply to Action Research. It must be knowledge directed, it must be calculated to produce new knowledge, or be intended to test, and maybe refute, existing knowledge. It must be systematically conducted. The chief questions to be addressed by the research must be unambiguously expressed. The methods of enquiry and analysis must be transparent. The data employed, and the observations made, must be fully and honestly recorded. And the whose must be published or otherwise exposed to critical examination by others. However, in one important respect Action Research is different from the other categories of research activity. Most Science research, at any rate, is planned and conducted in such a way as not to allow the enquiry processes to contaminate the phenomenon under investigation. The investigator tries not to interfere with the situation, or to influence the forces at work within it. He or she tries to ensure that personal values and expectations do not affect either observations or conclusions. In Action Research, however, the investigator is explicitly taking action in and on the real world in order to devise or test or shed light upon something. Sometimes, notably in educational research and medical research, the investigator is a significant actor in the human situation in which the action intervenes. In such circumstances, it is impossible to conduct the investigation on an interference-free and value-free and nonjudgmental basis. Consequentially, it is essential good practice for the Action Research investigator to make clear precisely what the intervention was, and exactly what was the theoretical, ideological and ethical position the investigator took up in making the intervention, observations and judgements. We have come across this consideration before, in connection with the subjective character of Humanities research, but it applies in Action Research even where the research methods employed are in other respects planned and exercised within the Science tradition.

Thus an important reservation has to be applied to research through practitioner action, as to all Action Research. It can hardly ever be objective, in the strict sense of the word. Moreover, Action Research is almost always 'situation-specific'. The term 'situation-specific' reminds us that, because Action Research is pursued through action in and on the real world, in all its complexity, its findings only reliably apply to the place, time, persons and circumstances in which that action took place. It is thus difficult and dangerous to generalize from action research findings. That is why orthodox scientists, in particular, are so suspicious of it. The Action Research investigator therefore has to keep it clear in his or her own mind that the investigation is necessarily situation-specific, and usually non-objective. He or she has to make it clear that the findings will only be generalisable to a very limited degree. Even so, Action Research findings are extremely valuable. They produce insights which might otherwise never be obtained. For a century or more they have provided case account material that has been extremely fruitful in the advancement of, for example, medical practice, agriculture, environmental studies and law. They have provided hypotheses for later testing in more generalisable Applied Research or Strategic Research programmes. Thus *research through practitioner action*, despite its being highly situation-specific, can advance practice and can provide material for the conduct of later, more generalisable, studies, provided the research is methodologically sound, the qualifications are clearly stated and the record is complete.

An investigator wishing to publish Action Research case study material often encounters problems of ethics, notably where individual people were part of the situation intervened in.

He or she may encounter conditions of commercial confidentiality, where industrial firms are concerned. In these circumstances, the investigator may have to find ways of publishing the research results for the benefit of other practitioners without revealing names, or other sensitive data. Professional scholars and researchers would probably still protest that such results were incapable of being verified, replicated or tested by others, and were therefore not recognisable in academic circles as a true contributions to knowledge.

Clearly, no matter whether a piece of research is *about* practice, or is conducted *for the purposes of* practitioner activity, or is conducted *through* practitioner activity, its status is determined by the conventions and standards of the class of research to which its procedures belong. Its reliability is determined by its methodology. In the case of *research for the purposes of* practitioner activity, however, there may be circumstances where it does not matter whether the research was well done or badly done, or whether the research results turned out to be true or false, or whether the findings, were situation-specific or generalisable. It may be sufficient to demonstrate that the practitioner outcome itself is satisfactory. In such a case, professional scholars and researchers would almost certainly protest that the investigation was, at best, Option Research, and at worst, not research at all, but mere speculation or exploration. The validation of the outcome of the practitioner work itself is another matter, of course, properly dealt with by field testing, or whatever.

To return to our intermediate question, it becomes clear that for academic recognition purposes a practitioner activity can rarely be recognised as in itself a research activity. One has to ask: Was the activity directed towards the acquisition of knowledge?

Was it systematically conducted? Were the findings explicit? Was the record of the activity 'transparent', in the sense that a later investigator could uncover the same information, replicate the procedures adopted, rehearse the argument conducted, and come to the same (or sufficiently similar) conclusions? Were the data employed, and the outcome arrived at, validated in appropriate ways? Were the findings knowledge rather than information? Was the knowledge transmissible to others? Only when the answers to all these questions are in the affirmative can a practitioner activity be classed as research.

Rewarded, Awarded or Assessed?

Let me summarise my argument. We have seen that not all research, however sound, qualifies the researcher for the award of an academic degree. There are many other kinds of reward for successful pieces of research; fees, patents, profit sharing, publication, fame. Those who share in, or promote, these rewards 'are all much more concerned with the outcome of the research than with the research methodology'. A researcher might, indeed, enjoy great profit from a research exercise that was, in fact, quite ineptly conducted, but had by chance achieved a demonstrably useful result.

I argued that practitioner activity can count as research if, and only if, it accords with the criteria of research. It must be knowledge directed, systematically conducted, unambiguously expressed. Its data and methods must be transparent and its knowledge outcome transmissible. But like all Action Research, research through practitioner action must be recognised as very probably non-objective and almost certainly situation-specific.

A research *degree* on the other hand, is primarily an acknowledgment of the competence of the person who conducted the research. For this reason, an examiner of a submission for a research degree is concerned much more with the soundness of the methodology than with the usefulness of the findings. Even a negative or empty result from research might still be rewarded with an academic degree if the methodology had been impeccable. This is because the identification of an empty field, or the refutation of an hypothesis, can nevertheless be a significant contribution to knowledge, and can demonstrate a satisfactory standard of research competence. Clearly, in every case of research conducted for the purpose of submitting for an academic degree, it is the quality of the research methodology that will be of paramount importance to the Examiners. Degree-worthiness is not quite the same as result-worthiness.

All this has been well understood for a long time.

What is new, perhaps, today, is the introduction of a new quality to be sought for in research: the elusive quality of Research Assessment Exercise-worthiness.

Examples of Action Research

Action Research is being carried out by Professor Philip Roberts and his colleagues at the Department of Design and Technology of Loughborough University of Technology. Amongst the projects which have been undertaken is an environmental project which focuses on a Leicestershire village. [...] The work was conducted by a group of fourteen year old children who looked into the environmental issues in their village. The work resulted in new play facilities being provided. Another project carried out by twelve year old children at a Leicestershire school involved a piece of graphically recorded data. Movements around a kitchen [...] were recorded with a view to the possibility of its better arrangement.

[...]

from: *Co-design, interdisciplinary journal of design*, (January 1995), 6-13

Alain Findeli

7 SEARCHING FOR DESIGN RESEARCH QUESTIONS: SOME CONCEPTUAL CLARIFICATIONS

1. Mode, purpose, subject-matter, and structure of the paper

The dominant tone of this paper is speculative and didactic. Speculative, since it does not rely on recent empirical research or field work; and didactic, in order to be in tune with the general framework of this conference and its presumed audience. The style has remained somehow that of the oral presentation. As to the purpose of the paper, it stems from the observed reactions to some previous published work on the same topic, namely design research methodology and education.[1] Some of the key concepts coined in these works seem to lack clarity, with the consequence that the resulting epistemological and methodological models suffer some misunderstanding and misinterpreting. For this reason, I considered it would not be superfluous to reframe these concepts and try to increase their intelligibility and, consequently, their usefulness in actual research situations.

The title of the paper directly mirrors the theme of our conference. In effect, this keynote lecture has been configured like a design proposal, i.e., as a hopefully adequate answer to a design brief; the brief in this case being the *Call for papers*, more specifically its "Why" section. From this section, I mainly retained the aim of being student-centred and the wish to promote "rigor in conceptualizing", especially in "formulating research questions", since "it is questions and ideas that give meanings and values to meticulously executed research".

The structure of the paper into two main parts is inscribed in its main title. First we will focus on the concept of design research, with the promise (made in Bern) that there will be no direct and explicit reference to Frayling's categories (!). The issue of what a research question is or should be will then be addressed, so that the following general questions may be answered: 1) Are design research questions very different from other discipline's research questions? And: 2) Is design research such a—reportedly—special case of research? In the conclusion, a general operational model of project-grounded research in design is presented.

2. Scope and stance: Another definition of design research

"Oh no, not another endless and useless discourse on the definition of design research!": Such may be the expected reaction to my proposal of redefining the field. This is fair enough, but the reason for such an apparently obstinate initiative is that I believe we, in our design research community, are using a somewhat restricted definition of the term. In other words, although I do agree with the members of the Board of International Research in Design that "It is no longer sufficient to merely indulge in either general or specific meta-discussions on methodologies or even on the fundamental question as to whether design is at all qualified to undertake research",[2] I also warn that, bearing with the metaphor of the pudding used by the authors, it is hazardous to look for a proof in the pudding by eating it if it is the wrong pudding that is being served. My remark is meant as a reminder that epistemological vigilance (e.g., to make sure we have the right pudding) is indeed always to the point, as it is—or should be—the rule in other areas of research.

Now why do I find it necessary to open this issue once again? Why am I not satisfied (I actually am) with the acknowledgment that "what is needed now is the publication of relevant results from design research",[3] or, to take another recent example, with the current state of the art of design research as reported in a book like *Design Research Now?*.[4] The reason, as will be argued shortly, is that we—the design research community—have built our collective design research enterprise on a misunderstanding. The statement of intentions and intellectual program of those we consider, with full right, the pioneers of design research were so convincingly spelled out that we have fared with them ever since in full trust, with an enthusiastic and almost uncontested unanimity.

Let me be more precise. There seems to be a common agreement, in our community, around Bruce Archer's definition of design research and Nigel Cross's search for a rigorous and compelling definition of his famous "designerly way(s) of knowing". As reported by Gui Bonsiepe and many others, in 1980 at the 'Design: Science: Method' conference, Bruce Archer mentioned in his talk the following definition: "Design Research is a systematic search for and acquisition of knowledge related to design and design activity",[5] The scientific validity of such a general statement can be checked by replacing "design" by any other discipline, for instance: "Economic research is a systematic search for and acquisition of knowledge related to economics and economic activity". If there is anything problematic with this definition, it lies with the definition of design one adopts. In this context, design is understood as the activity performed by designers.

The same holds, apparently, with Nigel Cross's "Designerly way(s) of knowing". Looking closer at this central concept, one finds out, first, that Cross alternatively refers to designerly ways of "knowing", "thinking" or "acting". As far as I know, he never discussed if he referred to the same epistemic process of "designerly" in all three cases, a task which would be of undeniable interest for the community. Nor does he explain why he alternately uses the plural or the singular. My intent here being mainly epistemological, I will proceed with the designerly way of thinking and try to characterize it further. For commodity reasons, I will stick to the singular, a conceptual generalization and risk I take all responsibility for.

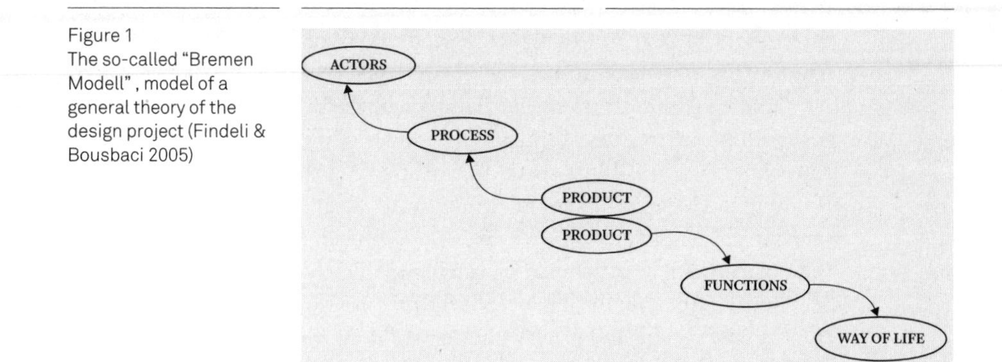

Figure 1
The so-called "Bremen Modell", model of a general theory of the design project (Findeli & Bousbaci 2005)

Indeed—and this is the important point—what Cross is interested in is the designerly way of thinking in design, i.e., in the specific logics and thought processes that designers adopt, individually or collectively, when doing design. In his view, the purpose of design research is then to observe, model, describe, theorize, and/or predict these processes in order, for instance, to show their specificity when compared to thought processes in other situations than design situations.

By no means do I mean thereby that Cross's intellectual and scientific program is irrelevant. There is plenty of evidence in the published literature and in the design studios that this endeavour has proven fruitful and valid. However, as we have shown in the article where the so-called "*Bremen Modell*" is introduced and discussed (Fig. 1),[6] the 'conception' part is only one of the two main moments or constituents of a design project, the 'reception' part being the other one (Fig. 2). When Cross uses the term "design", he only refers to the 'conception' side, whereas we consider that a model of the design act is incomplete if we do not address what happens to the project's output once it starts its life in the social world. In this regard, the opening up of the generic model of the design project to the user space is indeed one way of

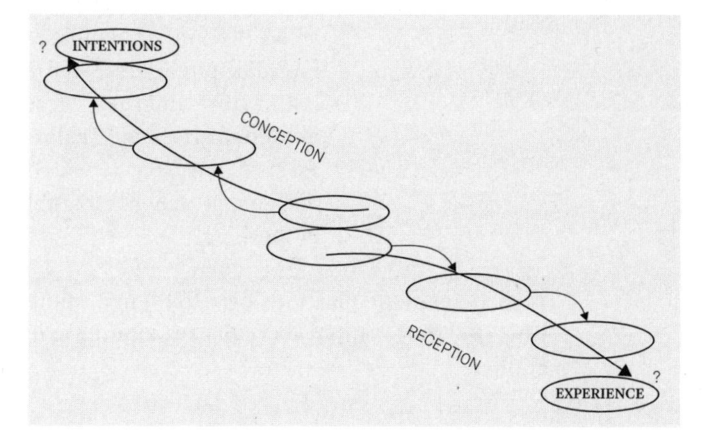

Figure 2
The Bremen Modell with emphasis on the two main moments or constituents of a design project

extending the scope of design research. As is witnessed by current research, such opening has proven fruitful.

What I contend however is that the scope of design research and of the designerly way of thinking can be extended much wider still, beyond the mere framework of design situations. I am interested in investigating the potential of a designerly way of thinking in research in general (not only in design), in the same way one might be interested in characterizing sociological, chemical, ethnological or others discipline's ways of thinking in research, i.e., when striving to know or understand the world. This amounts to consider design as a discipline on its own, capable of delivering valid and trustful knowledge about a part of the world considered as its specific field of knowledge. In such framework, our epistemological inquiry sets the task of determining what are the characteristics, the potential, and the blind spots of our designerly way of looking at the world and what the originality of the corresponding knowledge is. Thus our central question becomes: Does a designerly approach allow design researchers to increase or enrich the intelligibility of the world (or part of it) more or better than other disciplines?

Now that our task has been reset as clearly as possible, we must address the two following sets of questions, corresponding respectively to the scope and the stance of design research:

1) What is the proper subject-matter of design research? What is the part of the world design research may claim as being of its concern? To the knowledge and understanding of what phenomena is design research equipped to contribute?

2) How does design behold the world? Do design researchers observe, describe and interpret the world very differently from, for instance, ethnography, demography, economics or engineering researchers? More precisely still: Suppose design researchers are interested in the same phenomenon as the above disciplines (which may often be the case, especially in interdisciplinary research); in which way does their intellectual culture and their "designerly" approach colour the phenomenon? Does this "colouring" constitute a hindrance or an asset? Conversely, to which aspects of the phenomenon will design researchers remain blind, due to their designerly ways of thinking?

There are enough questions here to fill in a whole PhD project, so I will confine my answers to the essential at the risk of skipping some important and necessary justifications.

2.1 The scope or field of design research

It is generally accepted that the end or purpose of design is to improve or at least maintain the "habitability" of the world in all its dimensions: physical/material, psychological/emotional, spiritual/cultural/symbolic. The terminology may vary according to authors,[7] but the idea remains more or less the same. Habitability is best defined in systemic terms: it refers to the interface and interactions between individual or collective "inhabitants" of the world (i.e., all of us human beings) and the world in which we live (i.e., our natural and artificial environments, which includes the biocosm, technocosm, sociocosm and semiocosm). The discipline that studies these systemic relationships is human ecology: "Ecology is the science of relationships between living organisms and their environments. Human ecology is about relationships between people and their environment [...] [I]t is useful to think of human-environment interaction as interaction between the human social system and the rest of the ecosystem. The social system is everything about people, their population and the psychology and social organization that shape their behaviour."[8] Following such general definitions, who would deny that human ecology constitutes a core knowledge field for design?

The above conclusion brings us back to one of our previous questions: What distinguishes ecologists' and designers' claim that their central field of knowledge is the "relationships between people and their environment"? If there is no difference, then we should conclude that design research is or should be the same as research in human ecology. If there is a difference, then what is it?

In my view, the difference lies in two aspects. The first is anthropological (in the philosophical sense) and would deserve a longer discussion. Due to its rooting in biology, human ecology as a tendency to adopt a contextualist, determinist view of the human being; in this sense, human ecology is but an extension of animal ethology. For the purpose of design, the field of human ecology should be extended to the cultural and spiritual dimensions of human experience, consequently of the human-environment interactions, without for that matter neglecting the other dimensions. This is why I prefer to speak of a general human ecology. Keeping this important reservation in mind and using Bruce Archer's original phrasing as a template, we may redefine design research in the following terms:

> Design research is a systematic search for and acquisition of knowledge related to general human ecology.

The second aspect is epistemological. Design researchers' view of human ecology differs from ecologists' in what can be called their stance, i.e., in the way they look at the human-environment interactions. This will allow us to complete the above definition.

2.2 The stance or epistemological bent of design research

The aim of human ecologists is to construct a theory of human-environment interactions; their stance is descriptive and mainly analytical. Conversely, the aim of designers is to modify human-environment interactions, to transform them into preferred ones. Their stance is prescriptive and diagnostic. Indeed design researchers, being also trained as designers—a fundamental prerequisite—are endowed with the design intellectual culture: they not only look at what is going on in the world (descriptive stance), they look for what is going wrong in the world (diagnostic stance) in order, hopefully, to improve the situation. In other words, human ecologists consider the world as an object (of inquiry), whereas design researchers consider it as a project. Their epistemological stance may be characterized as projective.

The validity of the ecologists' descriptive/analytical stance derives from the grounding of their models, methods, and conceptual frameworks in their mother science, biology, the scientificity of which needs not to be assessed and asserted (any more). But what is the scientific validity of the normative, diagnostic, prescriptive, and projective position of design researchers, a stance which requires their subjective engagement? Are we not confronted here with one of the capital sins of scientific inquiry: lack of objectivity? What is indeed the value of a protocol which implies value judgments and includes the possibility that two different researchers will not yield the same conclusion?

Fortunately enough for design researchers, such epistemological scruples are not timely anymore in the scientific community. Recent developments in human and social sciences have dealt extensively with the issue of objectivity as a possible and desirable horizon in research. The interpretive or hermeneutic turn has shown that objectivity is not a relevant and fruitful criterion for research in those disciplines, and that rigorous inquiry is nevertheless possible without diving into extreme relativism or scepticism. On the other hand, the pragmatist epistemological tradition—where the engagement of the researcher is also required—may also be invoked to propose a robust epistemological framework for design research, not to mention action research (renamed "project-grounded research" in design research) as one of its incarnations in methodological applications.

As a consequence, with the warranty of careful and constant epistemological scrutiny, we may consider that a designerly way of looking at human-environment interactions, i.e., at human experience in terms of general human ecology, is not only a valid but also a valuable epistemological stance. In such conditions, design research has the potential of delivering original and relevant knowledge about the world, according to the following completed definition [9]:

> Design research is a systematic search for and acquisition of knowledge related to general human ecology considered from a designerly way of thinking, i.e., project-oriented, perspective.

3. Conclusion: Searching for research questions

If one adopts our redefinition of design research, the issue of the research question becomes more straightforward. For this purpose, the central distinction that needs to be made is between a research question and a design question. Our final model (Fig. 3c) will make clear this distinction by showing how, in a doctoral research situation, these two questions relate to each other:

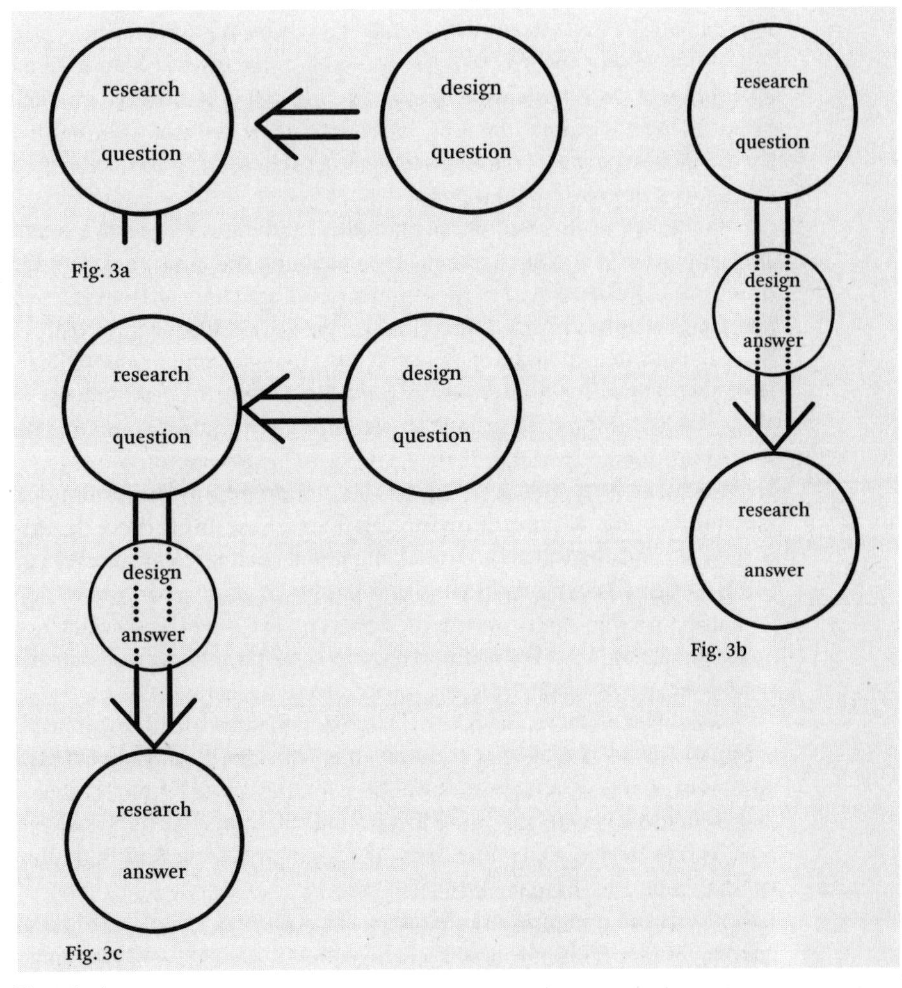

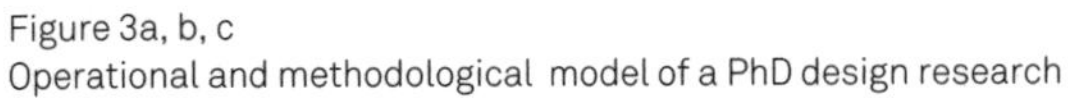
Figure 3a, b, c
Operational and methodological model of a PhD design research

A simple logical approach may already shed some light on this distinction. One may ask for instance if design questions constitute a subset of research questions or vice-versa, if research questions are a subset of design questions.[10] Or one may ask if research questions might be deducted from design questions (my viewpoint, with a reservation on "deducted") or vice-versa (my viewpoint also, with the same reservation). It is maybe wiser to ask oneself what the relationships between both realms of questioning are, in other words how different such questions "sound" or "taste", epistemologically and phenomenologically speaking of course. A quicker way to grasp the distinction is to look at how different the answers are to these questions. One notices for instance that design answers are presented in glossy design magazines with plenty of pictures and in sometimes very chic downtown galleries, whereas research answers are found in academic journals with as few pictures as possible in the typical grey literature and in—sometimes as trendy—academic conferences. More seriously, one could also compare the criteria used to evaluate both types of answers, an exercise that has been carried out quite extensively lately within our research community. At any rate, the distinction needs to be made in order not to confound or reduce a (design) research project with or to a design project.

A steady observation reveals that PhD candidates in design usually tackle their subject-matter in the form of a design question. The latter usually originates either from some unsatisfaction in their professional practice, or from the wish to deepen one aspect that has puzzled them in their professional education. This reflex is quite normal, but the next and important step that needs to be made then is to transform their design question into a research question. I posit that this is always possible since every design question raises, at least potentially, a bunch of more fundamental issues related to human experience in the world or, following our terminology, related to general human ecology. However, it is wrong and unfair to request from PhD candidates that they manage this transformation by themselves: this is the task, indeed the duty, of their supervisors, since in general, the intellectual and disciplinary knowledge and culture acquired by the candidates in their previous education and/or professional experience do not equip them with the necessary competence to switch from the realm of design questions to the realm of research questions. Only research experience and scholarship can provide the necessary intellectual mastery.

As hinted at above, the passage from design question to research question is not automatic or deductive. It is a matter of construction, of design. There are usually many potential research questions hidden in a design question, for the simple reason that design deals with the most banal of all phenomena: daily human experience. Who, except designers, is interested in such a prosaic and ordinary subject matter? For their inquiries, human and social sciences have a tendency to choose situations of exceptional and non-ordinary character: chess playing, social deviance, psychological distress, crime, economic crises, exotic cultural practices, etc. The daily life of ordinary humans has only recently raised some interest within academic circles (e.g., consumer studies, ethnography of contemporary societies, history of the present, popular culture, etc.). But this apparent banality of daily human experience conceals a rich complexity, well known by designers who are working in experience, service, or social design. Indeed every daily human activity (to go to work, to school, on vacation, to hospital, on retirement, shopping, etc., etc.) is an entanglement of various interrelated dimensions and values (economic, social, psychological, cultural, geographical, historical, technological, semiotic, etc.), each dimension being liable of a systematic inquiry and interpretation. The often proclaimed interdisciplinarity of design and design research is precisely a consequence of that.

An ideal design research question would thus be one that uncovers and emphasizes the complex interdisciplinarity of the specific anthropological experience that is at stake in a design question. In systemic, human ecology terms, we may consider each experience as the consequence of the interaction of two multilayered systems: the human (individual or collective) and his/her context or environment. A simple combinatory calculation shows that the scope of this complexity is out of reach of the usual conditions in which a PhD has to be carried out.[11] A choice has therefore to be made within all the possible research questions, an operation that requires a set of criteria. These are mainly circumstantial and situation specific: academic setting (large university with a vast choice of departments and disciplines or isolated design institution), personal experience of PhD candidate, scholarship of supervisor, surrounding research teams and laboratories, expectations of non-academic research partners (private or public), and of course personal inclination and intellectual project of candidate (and supervisor). These criteria need to be matched with the management constraints of the PhD research project: time, cost, availability of resources, nature of field work, etc. In short, the "search" problem of our title is not so much for the candidate to find a research question at all (there are way too many!), but to make sure to settle for a good research question.

Let us take a concrete example from the conference. In a previous presentation, a PhD candidate in architecture with obviously many years of professional experience explained us that, in social co-housing projects, one of the main obstacles to the realization of the architectural project was the mistrust arising between partners and stakeholders. He presented the problem as a very practical architectural question indeed, which has been answered (or not, in case of failure) diversely according to the singular situations. The presentation was very convincingly supported by slides and documented case studies by the speaker. Considering the wish of this architect to embark upon a PhD project, a possible starting recommendation could be to consider "trust" as his central concept of inquiry. The research question would then need to be worked out appropriately with the designerly (or architectural in this case) way of thinking perspective in mind. Concretely, this means that the idea is not to question what the concept of trust is or entails in general (a philosophical inquiry), or what mental processes are activated in a situation of trust (a cognitive psychological inquiry), or how the brain reacts to simulations of trust and mistrust (a neurobiological approach), or else what architectural historians and theoreticians have written on the concept of trust (provided they have), etc. Although all these aspects are indeed important and should ideally be addressed in the research, a more targeted (i.e., designerly) way of approaching the phenomenon could be with the following question: "Which facet of the general phenomenon of trust does an inquiry reveal that is actively engaged in an architectural project?". In other words, we take it for granted that the project-groundedness of the approach to the phenomenon of trust will contribute to the knowledge already provided by other disciplines that have studied the same phenomenon (law, ethics, religious studies, social psychology, anthropology, etc.). The title of the dissertation could thus read: "The contribution of architecture (or design) to a theory of trust".

The general model of this approach is what I have called project-grounded research in design, elsewhere usually called research through design. It derives from the pragmatist maxim (the "gospel of design research"): "If thou wantest to understand a phenomenon, put it into project".[12] The overall model is illustrated in figure 3c, and instructions read as follows:

Fig. 3a: First ask your supervisor to help you transform the design question into a research question. Then, remembering your PhD methodology seminar, conceive a research strategy corresponding to this question in order to reach a satisfactory research answer.

Fig. 3b: Select a research method where the designerly way of thinking is central to the research process to make sure you are in a design (i.e., not sociology, economics or engineering, etc.) research situation. For this purpose, use the design project as your research field as recommended in project-grounded research.

from: *Questions, Hypotheses & Conjectures - Discussions on Projects by Earl Stage and Senior Design Researchers*, Edited by Rosan Chow, Wolfgang Jonas & Gesche Joost, iUniverse Inc, 2010

NOTES

1 Reference is made here to the following publications: Findel, A., "Die projektgeleitete Forschung : eine Methode der Designforschung", in Michel, R. (ed.), *Erstes Designforschungssymposium,* Zurich: SwissDesignNetwork, 2005, 40-51; Findeli, A. & Bousbaci, R., "L'éclipse de l'objet dans les théories du projet en design", *The Design Journal* VIII, 3, 2005: 35-49 (with a long English abstract) ; Findeli, A. & Coste, A., "De la recherche-création à la recherche-projet : un cadre théorique et méthodologique pour la recherche architecturale", *Lieux Communs* 10, 2007:139-61 ; Findeli, A., Brouillet, D., Martin, S., Moineau, Ch., Tarrago, R., " Research Through Design and Transdisciplinarity: A Tentative Contribution to the Methodology of Design Research", *in* Minder, B. (ed.), *Focused-Current design research projects and methods,* Bern, SDN, 2008, 67-94.
2 BIRD, "Transition and Experience as Perspectives of Design Research", Foreword to Brandes,U., Stich S. and Wender M., *Design by Use,* Basel, Birkhäuser Verlag, 2009, 4.
3 Ibid., my emphasis.
4 Michel, R. (ed.), *Design Research Now,* Basel, Birkhäuser Verlag, 2007. For a review of this title, read D. Durling in *Design Studies,* 30, 2009: 111-12.
5 Bonsiepe, G., "The Uneasy Relationship between Design and Design Research", *in* Michel, R. (ed.), *Design Research Now, op. cit.*: In fact, Bonsiepe's report of Archer's paper is misquoted and misleading. On page 31 of the original paper, Archer writes that he finds the following definition of design research too narrow (notice the exact quote):"Design Research [with "a big D and R"] is systematic enquiry into the nature of design activity". He discusses instead two other possible definitions composed of the definitions of Design and design, on one hand, and of research ("with or without big R"), on the other hand. The first one he finds "impossibly broad": "Design Research is systematic enquiry whose goal is knowledge of, or in, the area of human experience, skill and understanding that reflects man's concern with the enhancement of order, utility, value and meaning in his habitat". He is "still uncomfortable with the vagueness of [the] focus" of the second one, even though it "seems to be a better description of the matter which design researchers are actually investigating": "Design Research is systematic enquiry whose goal is knowledge of, or in, the embodiment of configuration, composition, structure, purpose, value and meaning in man-made things and systems". Archer, B., "A View of the Nature of Design Research", in Jacques, R. & Powell, J. (eds), *Design, science, method,* proceedings of the 1980 Design Research Society Conference, Guildford, Westbury House, 1981, 30-47.
6 Findeli, A. & Bousbaci, R., "L'éclipse de l'objet dans les théories du projet en design", *The Design Journal, op. cit.* A long English abstract summing up the core argument is presented in the introduction of the article.
7 The concept of "habitability" has, to my knowledge, first been used in the early 80's by the Italians (Branzi, Manzini). Its origin is sometimes attributed to a famous text by Heidegger, *Bauen Wohnen Denken* (1951). Herbert Simon's terminology ("To transform a situation into a preferred one") is also quite popular within our community. One could also mention Manzini's most recent proposal that design should contribute to "Enable people to live as they like, while moving toward sustainability". Indeed, within such wide frameworks, more local purposes of design activity may be identified. Examples of such lists may be found in Bonsiepe, G., *op. cit.* (note 5 above), page 34 or Findeli, A. "De l'esthétique industrielle à l'éthique: les métamorphoses du design", *Informel* III, 2, summer 1990: 72.
8 The definition is taken from a standard textbook: Marten, G., *Human Ecology,* London & Sterling, Earthscan, 2001 (emphases in original text). Such a systemic

model has first been used to describe what I considered to be the *Urmodell* (in its Goethean phenomenological sense) of design activity. Findeli, A., "Design, les enjeux éthiques", *in* Prost, R. (ed.), *Sciences de la conception: perspectives théoriques et méthodologiques,* Paris, L'Harmattan, 1995, 247-73.

9 It is quite important to notice that the project-oriented perspective is not only required in the 'conception' constituent of design activity (Cross's program), but also in its 'reception' constituent. Wherefrom follows that users are not to be considered as mere "receptors" of the output of the design project (product, service, etc.), but as endowed with a project, namely the project of inhabiting the world in a meaningful, comfortable, functional, aesthetic, sustainable, etc. way. The terms 'reception' (borrowed from art history and theory) and "users" are somewhat misleading in this respect. In his own way, Bernard Stiegler makes a relentless and radical critique of the service economy and its concept of user. See his website: www.arsindustrialis.org.

10 Ranulph Glanville has addressed this problem as wittily as usual in various papers. His standpoint is that research situations are a special case of design situations.

11 Only considering a threefold anthropological model (physical, psychological and spiritual dimensions of the individual human being) and the previous fourfold partition of the environment (biosphere, technosphere, socio-political sphere, cultural/semiosphere), we are already in presence of twelve possible binary relationships to investigate, each one being in principle the specific domain of a scientific discipline. The complexity increases when ternary relationships are considered, since the subsystems are not independent from each other. A possible model may be found in *Human Ecology (op. cit.* note 8), page 2. Herbert Simon's somewhat behaviourist model has become an icon in our community. For an even more sophisticated systemic-anthropological approach, see Bodack, D., "Wie beurteile ich Architektur- und Designqualität?", *Mensch+Architektur* 41, April 2003: 2-15 (with English translation in annex).

12 See our "Research through Design and Transdisciplinarity", *op. cit.* note 1.

Ken Friedman

8 THEORY CONSTRUCTION IN DESIGN RESEARCH: CRITERIA: APPROACHES, AND METHODS

Design involves solving problems, creating something new, or transforming less desirable situations to preferred situations. To do this, designers must know how things work and why. Understanding how things work and why requires us to analyze and explain. This is the purpose of theory. The article outlines a framework for theory construction in design. This framework will clarify the meaning of theory and theorizing. It will explain the nature and uses of theory as a general concept. It will propose necessary and sufficient conditions for theory construction in design. Finally, it will outline potential areas for future inquiry in design theory.

Keywords: design research; design science; design theory; philosophy of design; theory construction

There comes a moment in the evolution of every field or discipline when central intellectual issues come into focus as the field and the discipline on which it rests shift from a rough, ambiguous territory to an arena of reasoned inquiry. At such a time, scholars, scientists, researchers, and their students begin to focus articulate attention on such issues as research methods, methodology (the comparative study of methods), philosophy, philosophy of science, and related issues in the metanarrative through which a research field takes shape. In many fields today, this also entails the articulate study of theory construction.

1. Definitions: design, research, theory

[...]

1.1. Defining research

The noun research means, 1: careful or diligent search, 2: studious inquiry or examination; especially: investigation or experimentation aimed at the discovery and interpretation of facts, revision of accepted theories or laws in the light of new facts, or practical application of such new or revised theories or laws, 3: the collecting of information about a particular subject: Merriam Webster's Collegiate Dictionary (page 1002)[1]; See also[2].

The transitive verb means 'to search or investigate exhaustively' or 'to do research for' something, and the intransitive verb means, 'to engage in research[3] (see also sources above).

The word research is closely linked to the word and concept of search. The prefix 're' came to this word from outside English. Rather than indicating the past as some have mistakenly suggested, it emphasizes and strengthens the core concept of search. The key meanings are 'to look into or over carefully or thoroughly in an effort to find or discover something, to read thoroughly, to look at as if to discover or penetrate intention or nature, to uncover, find, or come to know by inquiry or scrutiny, to make painstaking investigation or examination[4] (p 1059). Many aspects of design involve search and research together.

Basic research involves a search for general principles. These principles are abstracted and generalized to cover a variety of situations and cases. Basic research generates theory on several levels. This may involve macro level theories covering wide areas or fields, mid level theories covering specific ranges of issues or micro level theories focused on narrow questions. General principles often have broad application beyond their field of origin, and their generative nature sometimes gives them surprising power.

Applied research adapts the findings of basic research to classes of problems. It may also involve developing and testing theories for these classes of problems. Applied research tends to be mid level or micro level research. At the same time, applied research may develop or generate questions that become the subject of basic research.

Clinical research involves specific cases. Clinical research applies the findings of basic research and applied research to specific situations. It may also generate and test new questions, and it may test the findings of basic and applied research in a clinical situation. Clinical research may also develop or generate questions that become the subject of basic research or applied research.

Any of the three frames of research may generate questions for the other frames. Each may test the theories and findings of other kinds of research. Clinical research generally involves specific forms of professional engagement. In the flow of daily activity, most design practice is restricted to clinical research. There isn't time for anything else. Precisely because this is the case, senior designers increasingly need a sense of research issues with the background and experience to distinguish among classes and kinds of problems, likely alternative solutions, and a sense of the areas where creative intervention can make a difference.

In today's complex environment, a designer must identify problems, select appropriate goals, and realize solutions. Because so much design work takes place in teams, a senior designer may also be expected to assemble and lead a team to develop and implement solutions. Designers work on several levels. The designer is an analyst who discovers problems or who works with a problem in the light of a brief. The designer is a synthesist who helps to solve problems and a generalist who understands the range of talents that must be engaged to realize solutions. The designer is a leader who organizes teams when one range of talents is not enough. Moreover, the designer is a critic whose post-solution analysis considers whether the right problem has been solved. Each of these tasks may involve working with research questions. All of them involve interpreting or applying some aspect or element that research discloses.

Because a designer is a thinker whose job it is to move from thought to action, the designer uses capacities of mind to solve problems for clients in an appropriate and empathic way. In cases where the client is not the customer or end-user of the designer's work, the designer may also work to meet customer needs, testing design outcomes and following through on solutions.

This provides the first benefit of research training for the professional designer. Design practice is inevitably located in a specific, clinical situation. A broad understanding of general principles based on research gives the practicing designer a background stock of knowledge on which to draw. This stock of knowledge includes principles, facts, and theories. No single individual can master this comprehensive background stock of knowledge. Rather, this constitutes the knowledge of the field. This knowledge is embodied in the minds and working practices of millions of people. These people, their minds, and their practices, are distributed in the social and organizational memory of tens of thousands of organizations.

Even if one person could in theory master any major fraction of the general stock of knowledge, there would be little point in doing so. The general and comprehensive stock of design knowledge can never be used completely in any practical context. Good design solutions are always based on and embedded in specific problems. In Jens Bernsen's[5] memorable phrase, the problem comes first in design. Each problem implies partially new solutions located in a specific context. The continual interaction of design problems and design solutions generates the problematics and knowledge stock of the field in tandem.

Developing a comprehensive background through practice takes many years. In contrast, a solid foundation of design knowledge anchored in broad research traditions gives each practitioner the access to the cumulative results of many other minds and the overall experience of a far larger field.

In addition to those who shape research at the clinical edge of practice, there are other forms of research that serve the field and other kinds of researchers develop them. Research is a way of asking questions. All forms of research ask questions, basic, applied, and clinical. The different forms and levels of research ask questions in different ways.

Research asks questions in a systematic way. The systems vary by field and purpose. There are many kinds of research: hermeneutic, naturalistic inquiry, statistical, analytical, mathematical, physical, historical, sociological, ethnographic, ethnological, biological, medical, chemical, and many more. They draw on many methods and traditions. Each has its own foundations and values. All involve some form of systematic inquiry, and all involve a formal level of theorizing and inquiry beyond the specific research at hand.

Research is the 'methodical search for knowledge. Original research tackles new problems or checks previous findings. Rigorous research is the mark of science, technology, and the 'living' branches of the humanities'.[6] Exploration, investigation, and inquiry are partial synonyms for research.

Because design knowledge grows in part from practice, design knowledge and design research overlap. The practice of design is one foundation of design knowledge. Even though design knowledge arises in part from practice, however, it is not practice but systematic and methodical inquiry into practice—and other issues—that constitute design research, as distinct from practice itself. The elements of design knowledge begin in many sources, and practice is only one of them.

Critical thinking and systemic inquiry form the foundation of theory. Research offers us the tools that allow critical thinking and systemic inquiry to bring answers out of the field of action. It is theory and the models that theory provides through which we link what we know to what we do.

[...]

3. Theory construction problems in design research

Until recently, the field of design has been an adjunct to art and craft. With the transformation of design into an industrial discipline come responsibilities that the field of design studies has only recently begun to address.

Design is now becoming a generalizable discipline that may as readily be applied to processes, interfaces between media or information artifacts as to tools, clothing, furniture, or advertisements. To understand design as a discipline that can function within any of these frames means developing a general theory of design. This general theory should support application theories and operational programs. Moving from a general theory of design to the task of solving problems involves a significantly different mode of conceptualization and explicit knowledge management than adapting the tacit knowledge of individual design experience.

So far, most design theories involve clinical situations or micro-level grounded theories developed through induction. This is necessary, but it is not sufficient for the kinds of progress we need.

In the social sciences, grounded theory has developed into a robust and sophisticated system for generating theory across levels. These theories ultimately lead to larger ranges of understanding, and the literature of grounded theory is rich in discussions of theory construction and theoretical sensitivity.[7]

One of the deep problems in design research is the failure to develop grounded theory out of practice. Instead, designers often confuse practice with research. Instead of developing theory from practice through articulation and inductive inquiry, some designers simply argue that practice is research and practice-based research is, in itself, a form of theory construction. Design theory is not identical with the tacit knowledge of design practice. While tacit knowledge is important to all fields of practice, confusing tacit knowledge with general design knowledge involves a category confusion.

Michael Polanyi, who wrote the classic study of tacit knowledge,[8] distinguishes between tacit knowledge and theory construction. Where tacit knowledge is embodied and experiential knowledge, theory requires more. 'It seems to me,' he writes, 'that we have sound reason for ... considering theoretical knowledge more objective than immediate experience. ... A theory is something other than myself. It may be set out on paper as a system, of rules, and it is the more truly a theory the more completely it can be put down in such terms.[9]

Polanyi's[10] discussion of the Copernican Revolution adesses some of the significant themes also seen in Varian, Deming, and McNeil.[11] These address such concepts as descriptive richness, theory as a guide to discovery, and modeling. As a guide to theory construction, this is also linked to Herbert Blumer's idea of sensitizing concepts.[12] All of these possibilities require explicit knowledge, rendered articulate for shared communication and reflection.

Explicit and articulate statements are the basis of all theoretical activities, all theorizing, and all theory construction. This true of interpretive and hermeneutical traditions, psychological, historical, and sociological traditions, and it is as true of these as of quantitative research in chemistry, descriptive biology or research engineering, logistics, and axiomatic mathematics. The languages are different. However, only explicit articulation permits us to contrast theories and to share them. Only explicit articulation allows us to test, consider, or reflect on the theories we develop. For this reason, the misguided effort to link the reflective practice of design to design knowledge, and the misguided effort to propose tacit knowledge or direct making as a method of theory construction must inevitably be dead ends.

All knowledge, all science, all practice relies on a rich cycle of knowledge management that moves from tacit knowledge to explicit and back again. So far, design with its craft tradition has relied far more on tacit knowledge. It is now time to consider the explicit ways in which design theory can be built—and to recognize that without a body of theory-based knowledge, the design profession will not be prepared to meet the challenges that face designers in today's complex world.

4. Future directions

The goal of this article has been to examine criteria, approaches, and methods for theory construction in design research. To do this, I began with a foundation of definitions, using these to build a range of applicable concepts.

There is not enough room in one article to go beyond the general consideration of methods to a specific description of how to develop theory and build specific theories. This is a task for a future article.

Many avenues deserve exploration in the future. These include linking theory building to the perspectives of design science, proposing models of theory construction from other perspectives, generating theory from the practice of leading contemporary designers, and developing such basic tools as a bibliography of resources for theory construction and developing theoretical imagination and sensitivity.

Theory-rich design can be playful as well as disciplined. Theory-based design can be as playful and artistic as craft-based design, but only theory-based design is suited to the large-scale social and economic needs of the industrial age.

This systemic, theory-driven approach offers a level of robust understanding that becomes one foundation of effective practice. To reach from knowing to doing requires practice. To reach from doing to knowing requires the articulation and critical inquiry that leads a practitioner to reflective insight. W. Edwards Deming's experience in the applied industrial setting and the direct clinical setting confirms the value of theory to practice.

'Experience alone, without theory, teaches . . . nothing about what to do to improve quality and competitive position, nor how to do it' writes Deming.[13] in his critique of contemporary manufacturing. 'If experience alone would be a teacher, then one may well ask why are we in this predicament? Experience will answer a question, and a question comes from theory.'

It is not experience, but our interpretation and understanding of experience that leads to knowledge. Knowledge emerges from critical inquiry. Systematic or scientific knowledge arises from the theories that allow us to question and learn from the world around us. One of the attributes that distinguish the practice of a profession from the practice of an art is systematic knowledge. In exploring the dimensions of design as service, Nelson and Stolterman[14] distinguish it from art and science both. My view is that art and science both contribute to design. The paradigm of service unites them.

To serve successfully demands an ability to cause change toward desired goals. This, in turn, involves the ability to discern desirable goals and to create predictable—or reasonable—changes to reach them. Theory is a tool that allows us to conceptualize and realize this aspect of design. Research is the collection of methods that enable us to use the tool.

Some designers assert that theory-based design, with its emphasis on profound knowledge and intellectual achievement, robs design of its artistic depth. I disagree. I believe that a study of design based on profound knowledge embraces the empirical world of people and problems in a deeper way than purely self-generated artistry can do.

The world's population recently exceeded six billion people for the first time. Many people in today's world live under such constrained conditions that their needs for food, clothing, shelter, and material comfort are entirely unmet. For the rest, most needs can only be met by industrial production. Only when we are able to develop a comprehensive, sustainable industrial practice at cost-effective scale and scope will we be able to meet their needs. Design will never achieve this goal until it rests on all three legs of science, observation, theorizing, and experimenting to sort useful theories from the rest. To do this, design practice—and design research—require theory.

from: *Design Studies*, 2003, Vol 24, Issue 6, 507-522, Copyright Elsevier

REFERENCES

1 *Merriam–Webster's Collegiate Dictionary* 10th ed., Springfield, MA: Merriam–Webster, Inc (1993), 1002.
2 *Webster's Revised Unabridged Dictionary* (G & C. Merriam Co. 1913, edited by N Porter) 1224.; ARTFL (Project for American and French Research on the treasury of the French Language) Chicago: Divisions of the Humanities, University of Chicago (2002): http://humanities.uchicago.edu/forms_unrest/webster.form.html Date accessed: 2002 January 18; *Merriam–Webster's Collegiate Dictionary* online edition, *Encyclopedia Britannica,* inc., Chicago (2002) http://www.britannica.com/ Date accessed: 2002, January 21; *Cambridge Dictionaries online*, Cambridge: UK University Press, Cambridge, UK (1999): http://www.cup.cam.ac.uk/elt/dictionary/ Date accessed: 1999 November 21; Link Lexical FreeNet: Connected thesaurus, The Link Group at Carnegie Mellon University, Pittsburgh (1999) http://www.link.cs.cmu.edu/ Date accessed: 1999 November 21; Simpson J A and Weiner E S C (eds.) *Oxford English Dictionary* online ed. 1989, Oxford: Clarendon Press. Oxford University Press, (2002): http://dictionary.oed.com Date accessed: 2002 January 18; Brown, L (ed.) *The New Shorter Oxford English Dictionary* Oxford, UK: Clarendon Press and University Press (1993), 2558.
3 Ibid.,1002 and note 1.
4 See note 1, 1059.
5 J Bernsen, *Design. The Problems Comes First.* Copenhagen, Denmark: Danish Design Council (1986).
6 Bunge, M *The Dictionary of Philosophy*. New York: Prometheus Books, Amherst. (1999), 51.
7 B.G Glaser, *Advances in the Methodology of Grounded Theory. Theoretical Sensitivity,* Mill Valley, CA: The Sociology Press (1978); *Basics of Grounded Theory Analysis. Emergence versus Forcing,* Mill Valley, CA: The Sociology Press, Mill Valley (1992); B.G Glaser and A.A Strauss, *The Discovery of Grounded Theory. Strategies for Qualitative Research*, Chicago: Aldine Publishing (1967); A.A Strauss, *Qualitative Analysis for Social Scientists,* Cambridge: Cambridge University Press (1991); A.A Strauss and J Corbin, *Basics of Qualitative Research. Grounded Theory Procedures and techniques,* London: Sage (1990); A.A Strauss and J Corbin, "Grounded Theory Methodology": an overview. in: N.K Denzin and Y.S Lincoln, Editors, *Handbook of Qualitative Research*, Newbury Park, CA: Sage (1994).
8 M Poyani, *The Tacit Dimension*, Garden City, NY: Doubleday and Company (1966).
9 M Polanyi, Personal Knowledge, Chicago, IL: (1974), 4
10 Ibid., 3-9.
11 H.R.H Varian, 'How to build an economic model in your spare time.' in: M Szenberg, Editor, *Passion and Craft. Economists at Work*, University of Michigan Press, Ann Arbor: MI (1997); *W.E Deming, Out of the Crisis. Quality: Productivity and Competitive Position,* Cambridge: Cambridge University Press (1986); T.A Mautnerr, *Dictionary of Philosophy,* Oxford: Blackwell (1996); McNeil D.H, 'Reframing systemic paradigms for the art of learning', Conference of the American Society for Cybernetics (1993).
12 H Blumer, *Symbolic Interactionism. Perspective and Method,* Englewood Cliffs, NJ: Prentice-Hall, (1969); K Baugh, *The Methodology of Herbert Blumer. Critical Interpretation and Repair,* Cambridge: Cambridge University Press (1990); W.C Van den Hoonard, *Working with Sensitizing Concepts. Analytical Field Research,* Thousand Oaks, CA: Sage (1997).
13 W.E Deming, *Out of the Crisis: Quality, Productivity and Competitive Position*, Cambridge: Cambridge University Press (1986), 19.
14 Nelson H and Stolterman E 'Design as being in service.' in D Durling and K Friedman (eds.) *Doctoral Education in Design. Foundations for the Future*. Proceedings of the La Clusaz Conference, July 8–12, 2000. Staffordshire, UK: Staffordshire University Press (2000), 23–33.

John Chris Jones

9 A THEORY OF DESIGNING

The voice of reason asks ten questions:

1. Is there a theory of design?
2. What is the essential skill of designing—can it be described?
3. What are design methods?
4. How do you use them (design methods)?
5. Is scientific research useful in designing?
6. Is it scientifically possible to discover the nature of designing?
7. How to design complex systems?
8. How to solve or to avoid major problems created by the culture we create and inhabit?
9. How to redesign the designed culture?
10. How to teach this view of design—what are the principles?

to which the voice of intuition gives these spontaneous answers:

1. Is there a theory of design?

There are many but there is no accepted theoretical frame of reference that I know of, only the ability of the human mind and nervous system to react creatively, informedly and collectively in order to change things and situations (for the better, we hope). The basis is 'people'—not abstractions or reductions of what it is to be alive. To design is to trust people and the unknown, as we are, as it is. Henry Dreyfuss (1955), in his book *Designing for People*,was one of the first people to act on this idea.

2. What is the essential skill of designing? Can it be described?

I guess you mean the ability of people to arrive at new designs that are noticeably better (more imaginative, more inspiring, more useful, more profitable, less damaging) than what exists. A seeming magic.

This is at present a mystery and possibly it will remain one. The paradox is that human minds, both individually and in collaboration, can, when appropriately informed by design activity, or methods, somehow conceive of and evolve new things and situations that go beyond, and 'improve', what exists. Yet 'what exists' is their only ingredient. How is it done?

At its best creative activity works by identifying large parts of each person's bodymind with the new design itself—the distinction between designer and designed is thus weakened as the mind identifies with the world and the thing designed takes on the character of mind. Beyond that I can't say very much and don't feel the need to explain it—only to do it—as in breathing or in thinking or any in other activity!

It is more difficult than it looks: for instance, to compose this brief writing defeated me for several weeks—until I was rescued by the arrival of these rational questions ... Then I wrote my answers straight out, between midnight and three a.m. Why was that? I cannot say. When we do the things we do we often do not know how we do them. I did not anticipate that sentence, nor this.

3. What are design methods?

Techniques which enable people to design something, to go beyond their first ideas, to test their designs in use or simulated use, to collaborate in creative activity, to lead design groups and to teach and to learn designing. A method can be anything one does while designing: sketching alternative designs 'on the back of an envelope', calculating what are assumed to be the main parameters, formal brainstorming (and classification of the result), taking a rest, issuing a questionnaire, evaluating preliminary designs in 'affirmative groups' and, most importantly, observing and experiencing for oneself the use of existing or new designs (in real life or in simulations) ... A design method is any action whatever that the designers may decide is appropriate.

Test it! (said Christopher Alexander long ago) is the best design method there is.

But what are design methods you might still ask, hoping for a more theoretical definition or description. In reply I would say that the usefulness of a method (or the purpose of a whole design process, consisting of several methods in a chosen sequence or in parallel) is to provide an adequate way of 'listening to' the users, and to the world, in such a way that the new design becomes well fitted to people and to circumstances.

I sometimes think of designing as a meta-process, occurring before the product exists, that can predict enough of the future to ensure that the design can have the same quality of rightness that we see in natural organisms, in things that have evolved naturally, 'without design'.

I'd like to correct a misconception: when in the 1970s I criticised and appeared to leave design research it was not because design methods had become rigid tools that inhibited the imaginative skills of individual designers—it was because I was angry, and still am, at the 'inhumanity' of abstract design language and theories that are not alive to all of us as people, or to actual experience—and which threaten to reduce the reality of life to something less than human.

4. How do you use them (design methods)?

It is difficult to use formal design methods, but not impossible. One danger is that of drowning in the large amount of information that most methods generate—or otherwise losing one's way and losing confidence. In any creative process, what some of us call the intuition (or the imagination) must have priority. Reason (and science) must be used to support, not to destroy, this essential confidence and vision. Otherwise, the intuition, or creativeness, which does not perform to order, will 'fly out of the window'.

One essential is to reserve part of the design time and effort to 'navigation': the choice of methods and of evaluating them both by external results and by one's spontaneous feelings as the process proceeds. And in this, a key part of designing, we have to trust our informed intuitions, and the world, and other people, and our self-awareness (which is only available when we stop designing for a moment).

5. Is scientific research useful in designing?

When it comes to finding out how people experience and use existing or new designs, objective information is helpful, even essential—provided that the rigours of the laboratory and of scientific proof are not allowed to over-ride well-informed intuitions but are used to replace ill-informed ones. Scientific methods, like all others, must be subject to an intuitive meta-method of navigation, such as I described in the previous answer. Without that they can be unproductive or destructive of insight and of life. There is no objective way to prove what is right.

6. Is it scientifically possible to discover the nature of designing?

There is a huge difficulty here. It seems that scientific research is suited only to observation, inductive theory, and experiment in relation to things that exist and that are separate from the people who are doing the research. Studying creative thought processes, such as design, is I think better done directly by introspection, by empathy and by conversation, etc.

The underlying difficulty of studying design is that it is concerned with the whole of something and 'the whole' is not an objective reality—it is a fluctuating scheme or state of the bodymind, perhaps more akin to religious inspiration than to science or to technology (William James 1901-2). One can be totally involved. And what I am calling the bodymind includes not only the brain and body but one's conceptions of both body and of world. These are not external objects but they are surely realities.

(At last I feel that this writing is flowing well—the complexity of life is at last entering into the argument!)

7. How to design a complex system?

I've read much about complexity, and I am more than aware that complex design problems—such as urban traffic, or the provision of medical care to all who need it—are unsolved in any city or region. Adequate designs for such situations call for social changes that are presently resisted and are perhaps beyond us, as yet (Jones 2000).

I am also aware that it is not possible to measure complexity—all entities are as complex as each other—and the definition of what is a 'thing', 'system' or an 'entity' is essentially arbitrary, as is any word. Word is a word and so is this. The words of a theory of design are also designs—there is no way to step outside of 'oneself being the world' but unfortunately the teaching of science as objective method gives the impression that one can . . . (Is that it?)

8. How to solve or to avoid major problems created by the culture we create and inhabit?

I suppose you are thinking of problems such as global warming (and to know how much of it is or is not the result of human activity), urban terrorism, global poverty, globalisation (is it good or is it bad for people and for the natural environment?), etc. *The Encyclopedia of World Problems*, 1994, lists several thousand such problems, and strategies for solving them—but in practice they are mostly unsolved.

I believe that the way to tackle these is to change (with the help of computernets) to a non-specialist culture in which computers are the specialists and all people are enabled to play creative parts in the continual redesign of the industrial culture as we live it. I call this (as yet unclarified) idea 'creative democracy' (Jones 2000). It is intended to replace representational democracy by continual voting and by responsible action by everyone (in place of work as a specialist, paid to attend only to parts of the whole but not to it).

9. How to redesign the designed culture?

It is evident that these large design problems of the time cannot be solved without changing the culture so that everyone plays a part in re-creating it—in a continual process I call 'designing-as-living' (Jones 1993). If this is to happen it will I think be necessary for design and design research to be unified into a single process of which they are complementary parts. That is I believe the first move, from which others could follow.

But how could that be done?

I see three changes as being necessary.

Firstly, designers would be paid not to design only individual products but the whole experience of each of us as we endure or enjoy the succession of products that we encounter each day—and to foresee and to design the combined effects of new designs on their surroundings.

Secondly, professional design researchers would be paid to seek answers to any questions that arise in the course of designing 'the whole experience' of modern life and the effects upon the environment (which of course includes 'the mind' and all artifacts, seen as products of natural beings). For this, 'objective' research methods are I think inadequate.

Thirdly, designers and researchers would be required to relate their activities firstly by practicing and learning each other's skills, and then by integrating them within their own minds and in ways indicated by my answers to the other questions.

A fourth stage is to involve all the people who experience the effects of new designs in the design process. Eventually this could lead to the ending of design and research as specialised professions as everyone becomes a designer-researcher, in part, working through software in which the describable parts of each activity are embedded in software accessible to all ... But this is in itself a large and complex design that is perhaps beyond the scope of these questions and this discussion? If there is an answer I expect it to resemble the operation of the 'open source movement' in the spontaneous and collective design of the internet (Naughton 2000)

10. How to teach this view of design—what are the principles?

I don't see this as a set of principles but I do have some suggestions arising from experience:

Firstly that the teacher (if there is one) carries out the same design tasks that he or she sets for the students—and is willing to discuss the doubts, and the often fragile imaginings, that underlie his or her creative work. This should encourage the students to do likewise.

Next, both teacher and students, collectively and consciously, choose appropriate methods to suit both the design problem and their varying states of mind as designer-researchers (or artists of science?). And they change methods at intervals, accordingly.

Thirdly, each designer begins to learn how to transfer the creative task, or opportunity, to the people who will use or inhabit (or otherwise experience or suffer) the design, using appropriate methods (such as designing only parts, not wholes—like words, bricks or computers)—from which people can improvise imaginative and flexible designs according to unpredictable situations arising.

I could make other suggestions but perhaps that is enough to show how unprincipled and flexible is designing, at its imaginative best—as a joy, not a problem—as it changes the variables!

(Intuition thanks reason for her questions—without which this theory would not have appeared.)

from: softopia: my public writing place at www.softopia.demon.co.uk/2.2/theory_of_designing.html

REFERENCES

Dreyfuss, Henry, (1955) *Designing for People*, out of print but a second-hand copy can sometimes be found at www.amazon.com.

The Encyclopedia of World Problems and Human Potential, (1994–5) volumes 1–4, ISBN 3–598–11165–7, 4th edition, edited by the Union of International Associations, Brussels (initiated by Anthony Judge). There are printed and online versions:http://www.uia.org/uiapubs/pubency.htm and http://www.uia.org/data.htm

James, William (1901–2) *The Varieties of Religious Experience: a study of human nature*, Gifford Lectures, Edinburgh, and in several editions including Fontana Library, Collins 1960.

Jones, John Chris, (1993) 'Designing as Living?', *Information Design Journal*, volume 7, number 2: 142–148.

Creative democracy is tenatively described in my recent book *the internet and everyone*, John Chris Jones, 2000—a summary of the relevant pages appears on my website at http://www.softopia.demon.co.uk/2.2/creative_democracy.html Descriptions and proposed solutions to the urban traffic problem and the future of medicare appear on pages 46–50 and 491–501 respectively.

Naughton, John (2000) A brief history of the future: the origins of the internet, London: Phoenix Paperback—it includes the best account I know of the open source movement.

Simon Grand

RESEARCH AS DESIGN:
Promising Strategies and Possible Futures

Prologue: Current Perspectives on Design Research

In current debates on design research, we observe a tendency to discuss the qualities and possibilities of design research based on a more or less taken-for-granted concept of (scientific) research and design. Insights, methods, processes and presentation modes from scientific research are transferred into the field of design and used to describe, understand, explain, and perform design as a scientific practice; scientific communities, scholarly journals, disciplinary agendas and interdisciplinary collaborations, and PhD programs, have all been developed in recent decades. In turn, design practices and "designerly" knowledge are transferred into scientific research: we observe a growing interest in better understanding the importance of images, artifacts and interfaces, as well as creation and design processes, in scientific research. In parallel, the basic difference between scientific research and design practice is emphasized, arguing that they are incommensurable, losing their respective qualities and idiosyncrasies if related to each other: in particular, design practices are subverting, challenging and criticizing scientific practices. In most cases, a certain basic understanding of (scientific) research (including the importance of methods, the ideal of intersubjectivity, the existence of a scientific community, the grounding in data) and design (as an inherently intuitive and imaginative practice, which focuses on the creation of new artifacts and images) is necessarily assumed in these discussions.

The multiplicity of design areas (from industrial design to scenography and fashion design), and the multiplicity of scientific disciplines (from cultural theory to life sciences and sociology), which actually and potentially are related, can lead to a heterogeneous, often rather confusing, situation. We can identify different strategies used to approach this heterogeneity (see Jonas 2011 in this volume): some designers and design researchers *map this heterogeneity of design research*, as a way to structure the field and to position themselves within this field (Laurel 2003; Sanders 2006; Jonas 2007); some designers and design researchers see research as *an opportunity to finance their projects*, interpreting design research as one way of generating and attracting financial resources; some designers and design researchers *take a methodological, theoretical, or epistemological perspective*, launching their own explicit or implicit conceptualization of design research (Glanville 1988; Simon 1996; Cross 2006; Krippendorff 2007); some designers and design researchers refer to a particular perspective or research approach, in order to *subvert, criticize, and challenge their own particular practice as a designer*, or a particular research practice, or any other social, cultural, or political area (Dunne 1999; Baur 2009). Overall, "design research" does not exist as a defined and well-structured discipline or field of activities, and probably never will. Paraphrasing Bruno Latour's essay on recalling actor-network theory, one can thus argue that three things are particularly problematic in most discussions of design research: the word "design", the word "research", and the two words in combination (see Latour 1999, 15).

Overall, this essay argues for a *fourth avenue for design research* (besides research "about" / "into", "for", and "through" / "by" design, as it is typically discussed, see Jonas 2011 in this volume), by interpreting *research "as" design* (Glanville 1988). Thus we will not add to the attempts to structure, define, conceptualize, map, and clarify the messy and controversial field of design research. We begin instead with the premise that it is *of central importance for design research to remain messy and controversial*. We focus on exploring promising strategies and future possibilities for design research: instead of defining design research in terms of what it is, we focus on *how we can re-invent possible futures for design research*. This implies a particular understanding of design research (which we will call "Design Fiction"); it also refers to an often implicit, sometimes unconscious, seldom reflected, founding paradigm in the field, that identifies "design" as transcending current realities and being oriented towards possible alternative futures (which could also be called "Entwurfsforschung": Bonsiepe 2007; Findeli 2007). We structure the essay as follows. In the first part (*Design as Practice and Research: Basic Premises*), we discuss some often cited qualities of design practice, so as to deduce promising strategies and future possibilities for design research. In the second part (*Research as Design: Insights from the Science Studies*), we discuss insights into the practice of scientific research, in the light of the science studies, in order to deduce promising strategies and future possibilities for research in general, and for design research in particular. In the third part (*Design Fiction: Promising Strategies for Design Research*), we condense our insights into those research strategies which we find particularly promising for design research. We end with a short epilogue (*Possible Futures for Design Research*).

Part 1: Design as Practice and Research: Basic Premises

In the current debates on design research, we find a series of qualities that are often cited as being inherent to and characteristic of "designerly ways of knowing" (Cross 2006; Krippendorff 2007). In this first part of the essay, we explore in what ways these qualities can be seen as inherent to and characteristic of (scientific) research. *We argue that particular design practices can be interpreted as research practices*. From there, we explore how design as outcome, method, process, and activity are intertwined, forming design as a practice (Schatzki, Knorr Cetina & von Savigny 2001). We argue that it is important in the context of design research for design practice to reflect, explore, exploit, and advance these qualities, in order to qualify as design research, because these qualities cannot just be taken for granted. Furthermore, we argue that it is not these qualities *per se*, but their enactment and performance, that is essential for the activity to qualify as design research. This implies that *design can be seen as research, if it explicitly attempts to qualify as design research, with respect to some essential quality criteria*.

We do not argue that these quality criteria are objectively set or scientifically given; rather, we argue that each design practice attempting to qualify as design research must in one way or another explicitly or implicitly refer to such quality criteria (Elkana 1986; Knorr Cetina 1999). *Without controversial and explicit debates concerning these qualities, referencing something like "design research" does not make any sense (in fact, the same holds for "research" and for "design" in general).* We introduce four central dimensions of design practice, with respect to which such quality criteria are particularly important.

1. Designing New Artifacts

Designing new artifacts can definitely be identified as a central characteristic of design. In the perspective of current design research debates, it is argued in particular that designing objects is essentially related to an anthropological premise of human existence: by designing, we can transcend ourselves (Gehlen 1961; 2004; Honneth & Joas 1980; Flusser 1996; 2008), thereby extending our possibilities in space and time, through the creation, enactment, and dissemination of physical artifacts. Today, this line of thought is extended into the study of the *multiple ways in which new media and technologies make it possible for us to extend our capabilities, sensibilities, and modes of interaction as human beings* (Brooks 2002). Furthermore, building new artifacts and objects is a particularly promising approach for understanding how things work, how they interact, and how they are used in social and cultural interactions. Overall, this means that *design practice as the creation of new artifacts can be seen as a particularly promising research strategy, also called synthetic method,* in the field of artificial intelligence and the cognitive sciences (Pfeifer & Bongard 2007): *analyzing, understanding and explaining by building.*

2. Designing New Images

With the advent of the so-called "pictorial turn", also called the "iconic turn" (Maar & Burda 2004), the importance of images for every aspect of our economic, technical, cultural, social, and political life is made manifest. Taken-for-granted approaches to researching art and design from the perspectives of art history and cultural theory are extended by research on pictures, images, representations, and visualizations in scientific laboratories (Galison 1997) and scientific exhibitions, as well as research on *image-making in such diverse and interrelated areas as science, religion, and art* (Latour & Weibel 2002). Thus the historical reconstruction of image production, along with the interpretation of images and their impact, constitutes only one contribution of iconic research. For design research, it is particularly important to research the *design practices involved in creating, collectivizing, producing, and distributing pictures and images.* Therefore, it is argued that visualization is not only about representing the world as it could be seen, but also about experimenting, so as to *imagine, visualize, and project alternative world views* (Boehm 2007): *making actual and possible worlds visible, and thus making the processes of their creation comprehensible.*

3. Designing New Interfaces

Given the increasing complexity of technological, cultural, and social systems, it is difficult or impossible to fully understand these systems and domains, which implies that we are increasingly forced to interact with these systems as black boxes (Latour 1999). This means that we must trust not only the artifacts and images, but also the information and knowledge surrounding us (Sloterdijk 2006). Designing interfaces with such artifacts and knowledge domains allows us to act and decide *as if* we understand, without actually understanding (Bolz 2006). Design is a way of *ensuring agency in the face of increasing complexity, uncertainty, and ambiguity*, an important area for design in the context of research, discussed under diverse labels, including interface and interaction design, and knowledge and information design. Thus it is important to recognize that ensuring functionality and creating meaning are closely related (Sloterdijk 2009): "After several centuries with new forms of life, we understand that human beings not only live in 'material conditions,' but also in 'symbolic immune systems'": *creating material and symbolic, tangible and intangible interfaces to exploring knowledge domains and technological systems.*

4. Designing New Usages

We thus have to consider that artifacts, images, and interfaces are never completely predetermined, but are co-designed in their use by the users (Orlikowski 2002, 2000). The use of artifacts and technologies must be seen as an important part of the creation and innovation process itself, either simply situationally (Kelley 2001), or systematically through user communities that are explicitly involved in creating, establishing, and designing new artifacts and technological systems (von Hippel 2005). As a consequence, *usage must be increasingly interpreted as an important area of research in itself*, either through ethnographic observation, or through collective laboratories. In this way, the material world becomes intertwined with the symbolic world in multiple ways. Developing new artifacts and technologies, images, and interfaces enables us to "... transcend the creation of an object purely to satisfy a function and necessity. Each object represents a tendency, proposal, and indication of progress with a more cultural resonance ..." (Utterback et al. 2006: 166): *understanding by enacting and simulating new usages and interactions.*

Designing new artifacts, images, interfaces, and usages not only implies that design as practice focuses on particular phenomena, but also that it is possible to identify specific "designerly" ways of approaching these phenomena. In the next section, we discuss four important qualities of a "designerly" way of knowing.

5. Design Implies Creation

Design is typically identified with creation: one central concern of design is the conception and realization of new things. Currently, this characteristic of design practices is discussed in the context of the so-called "creative industries" (Florida 2002). We also observe a growing interest in "design thinking" as an important source of innovation and change (Brown 2005; 2008; 2009). However, instead of just taking it for granted that design practices will "inherently" lead to the creation of new things and thus to innovation, and rather than identifying the "creative industries" with the actual and potential relevance of design practices for creation and innovation, it is important to explore *how and why design practices lead to the creation of new things and to innovation*, thus opening and unpacking the black box of creativity and creation. Furthermore, it is important to *better understand the criteria and dimensions with respect to which we explore new things and innovation as being "new"* (Verganti 2009). It is important for design research to *identify the settings, processes, methods, and approaches that make it possible for us to explore design as creation "at the edge".*

6. Design Implies Intention

Currently, we observe a rapid expansion of the areas in which the term "design" is used to describe intentional imagination and the construction of something new: from urban design to design principles in artificial intelligence (Pfeifer & Bongard 2007), from the design of a new technological system to managing as designing (Boland & Collopy 2004), from the design of business models to genetic engineering as design (Nowotny & Testa 2009), and from designer drugs to knowledge design. Thus design is associated with the *artificial, synthetic, intentional creation of something new, typically related to new artifacts and technologies in almost any economic, social, and cultural area*. However, while the term "design" suggests the possibility of intentionally creating synthetic new things, in a proactive and controlled way, we know from the fields of the science and technology studies just how *complex and distributed intentionality and agency are, especially in the context of creation, innovation, and design processes* (Bijker, Hughes & Pinch 1989). It is thus important for design research to better *understand the processes underlying not only the formation and emergence, but also the simulation and imagination, of intentions.*

7. Design Implies Materiality

"Sketching" is identified as a central design activity, operating at the interface between the future and the present, the possible and the actual, the imaginative and the real (Gänshirt 2007). Sketching takes place in multiple media, which allows the exploring and materializing of new ideas, concepts, strategies, and perspectives in multiple ways. The use of particular media for sketching has an impact on the specific development, evaluation, and realization of these ideas, concepts, strategies, and perspectives (Gänshirt 2007): sketching takes place in diverse media and related practices, and includes *drawings, images, narratives, models, collages, simulations, and mood boards.* Furthermore, sketching *relates tools, methods, experiences, imagination, and activities in complex ways that are called "practices"* in recent cultural and social theory (Schatzki, Knorr Cetina & von Savigny 2001). More recently, the focus has been shifting towards studying *how such practices are actually performed* (Franck & Franck 2008); in order to better *understand the particular qualities of materialization in the design process and as a designed outcome.*

8. Design Implies Process

We design by imagining possible future worlds, while at the same time referring to the world as it is now. In this perspective, *design implies a continuous oscillation between the present and the future, the actual and the possible, the real and the potential* (Hernes 2008). Thus the mode of referring to the present and the approach towards the future are related, leading to a multitude of design strategies: dis-assembling the taken-for-granted in order to re-assemble possible alternatives (Latour 2005); criticizing the present in order to explore ways of transforming the world (Boltanski 2009); re-interpreting the past in order to explore new approaches (Tsoukas 1991); creative de(con)struction as an essential procedure for innovation (Schumpeter 1942); subverting dominant discourses in order to establish alternative ways of talking (Foucault 1971). Obviously, the close relationship between referring to the present (and the past) and imagining possible futures implies that design as imagination and project(ion) (Findeli 2007) must be seen as a *particular interpretation of the world as a dynamic process leading into the open, uncertain ambiguity of future possibilities.*

Design is inherently contingent. Whatever we design could be otherwise; it is thus assertive and fragile, and must be stabilized in order to transcend the situational in time and space (Grand 2009) and thus attain broader impact. As a consequence, it is of central importance to better understand the *multiple strategies, mechanisms, procedures, and approaches through which design is engaging multiple actors, people and artifacts, systems and technologies, discourses and communities* (Latour 1999), in the context of design practice as well as in the context of design research. Furthermore, it is important to *better understand how these strategies, procedures, and methods are carried out in particular processes*, how they act upon and shape design as outcome, design as process, design as method, and design as activity, and thus how they are intertwined, and thereby inform design as practice. Finally, it is important to consider the particular criteria and *dimensions of "quality", that have to be fulfilled in order to qualify as "good" practice or as "good" research beyond contingent self-declarations.*

Part 2: Research as Design: Insights from the Science Studies

In the science studies, we observe a growing interest in interpreting scientific research as a constructive and creative practice (Knorr Cetina 1999). Scientific research itself is a particular way of acting upon and shaping reality, and a way to prepare for possible futures. If we take this seriously, we can identify two implications: first, what we accept as scientific knowledge at a particular point in time is emerging from ongoing controversies among multiple parties (Latour 1999); second, what we discuss as "designerly" ways of knowing are not inherently either scientific or non-scientific, but *must be seen as particular perspectives and possibilities for scientific research, which over time either are accepted as scientific* (Foucault 1971) or remain dissident (Krippendorff 2007). The science studies are particularly inspirational in opening up new perspectives on scientific research and on the production of (scientific) knowledge, in scientific research generally, but also in design research more specifically. At the same time, the science studies are actually and potentially relating to the phenomena and qualities that we have discussed as being characteristic of design practice and design research.

Scientific Research as a Controversial Process of Knowledge Production

Overall, it can be argued that the science studies describe scientific research as being less orderly and aligned than we expect, as it is often seen in an idealized perspective from the outside. Scientific research in the eyes of the science studies must be seen as an *open, uncertain, ambiguous, and controversial process of knowledge production*, in which multiple perspectives, concepts, methods, theories, and questions ... are competing for attention, legitimacy, resources, and self-evidence. The following three positions are particularly insightful in terms of our discussion of design research as a (scientific) research domain (see the reader section for excerpts from the three main underlying book chapters and scholarly articles).

1. "Trading Zones" in Scientific Research

It can be argued that "*... science is disunified—and against our first intuitions—it is precisely the disunification of science that underpins its strength and stability ...*" (Galison 1997, 137). The debates among different, heterogeneous perspectives and competing approaches in a scientific research area are essential for the development, enrichment, and advancement of scientific research. These debates continuously "push the envelope" of scientific research, with respect to such diverse things as new phenomena, new methods, new theories, new questions, and new technical possibilities. For design research, rather than trying to establish collectively accepted and well-structured definitions of the field, it is much more important to foster lively debates and controversies (Elkana 1986). On the one hand, this implies that different theories, methods, techniques, paradigmatic phenomena, and guiding research questions must co-exist and compete against each other in a research area; on the other hand, it implies that these constellations and assemblages of theories, methods, techniques, phenomena, and questions are always shifting and changing over time. This shifting and debating, controversy and competition, takes place in what may be called "trading zones": "*... I intend the term trading zone to be taken seriously, as a social and intellectual mortar binding together the dis-unified traditions of experimenting, theorizing, and instrument building ...* " (Galison 1997, 145 ff.). In particular, this implies that there is no obvious alignment between theory and practice in scientific research, and thus also in design research; the two have to be disassembled and reassembled in varying ways over time: *establishing trading zones and discussion platforms that allow multiple controversies and intense competition among alternative approaches to research and design is of central importance for design research.*

2. The New Production of Knowledge in Mode 2 Research

In a similar line of argumentation, scientific research is described as changing its dominant mode of knowledge production over time. It is argued that research is currently shifting from a disciplinary, cognitive constellation towards a broader, transdisciplinary, social and economic constellation. This research mode is not primarily characterized by constant debate within a particular scientific discipline, but rather encourages debate between disciplines, among social and cultural, economic, and technological fields that are all interrelated in multiple ways: "*... the relevant contrast here is between problem solving which is carried out following the codes of practice relevant to a particular discipline (mode 1) and problem solving which is organized around a particular application (mode 2),*" which is transdisciplinary, bringing together multiple knowledge domains and disciplinary perspectives (Gibbons, Limoges, Novotny, Schwartzman, Scott & Trow 1994, 3 ff.). To make use of the skills and experience levels, concepts, and expectations that come together in such settings, innovative forms of structuring scientific research are required. Furthermore, mode 2 research requires societal accountability and public justification, since "*... working in the context of application increases the sensitivity of scientists and technologists to the broader implications of what they are doing ...*" (Gibbons et al. 1994, 7). The areas, activities, and problems that relate to (scientific) research in one way or another are thus expanding: "*... in parallel with this vast expansion in supply (of research) has been the expansion of the demand for specialist knowledge of all kinds ...*" (Gibbons *et al.* 1994, 11). *Design practice and design research are inherently involved in mode 2 research, while at the same time constituting in themselves promising areas for transdisciplinary research.*

3. Translation Processes in Research

We can describe how scientific knowledge is created, as well as how scientific controversies discuss and define the scientific value and relevance of knowledge, involving heterogeneous actors, activities, artifacts, and actions (Callon 1986). The sociology of translation identifies four fundamental processes in this context (see also Latour 1999). First, any scientific debate in one way or another leads to the *problematization* of existing knowledge, including particular phenomena and their interpretations, theories and related methods, questions and established answers. Such debate discusses the contingency of assumed realities and perspectives, and launches important puzzles and alternative views. Second, processes of *interessement* clarify how existing and alternative perspectives relate to each other, identifying and establishing common interests. Third, processes of *enrolement* ensure that these other actors have particular roles in the new perspective, as a precondition for the fourth process, *mobilization*, through which they become involved in shared activities, alliances, and programs. These processes are thus open and uncertain: "*... the notion of translation emphasizes the continuity of the displacements and transformations which occur in this story ... translation are a process before it is a result ...*" (Callon 1986, 81). Translation processes are thus characterized by the dynamic interplay between processes of opening up, criticizing, subverting, and challenging current scientific knowledge, and processes of closing down, stabilizing, clarifying, and ending controversies (Latour 1999). This perspective is important to design practice and design research, as it *explicitly focuses on the co-creation of worldviews, the world itself, relevant artifacts and actors, and images and interfaces, as well as how their mutual interdependence is challenged or stabilized in the research process.*

Any discussion of design research and design practice benefits from the science studies as they identify a series of patterns in scientific research and knowledge production that are also relevant here: the close interplay between artifacts, images, interfaces, and usages in design practice can be seen to parallel the interplay between ideas, methods, phenomena, and theories in the scientific "*trading zone*", including interactions with other knowledge domains (economic, technological, societal, cultural) and related actors and artifacts (as discussed in the context of *mode 2 research*). Thus they are related and assembled in more or less self-evident, taken-for-granted ways, which result from translation processes, and which at the same time can be problematized again through *translation processes*.

Scientific Research as Experimentation within and beyond Laboratories

An exemplary context for scientific research, as discussed in the science studies, is the laboratory, and the related exemplary process for scientific research is seen in the process of experimentation. The following three positions offer insight into design research as a scientific research domain, each for particular reasons. First, we can focus on the laboratory as the special and material, but also cultural and epistemic, space for scientific and design research. Second, we can focus on experimentation processes in scientific and design research by exploring their particular preconditions, characteristics, and dynamics. Third, we will have to discuss an emerging new perspective that observes an extension of current experimental settings, going beyond the close boundaries of scientific research, and thus also opening up towards design practice and design research, and also going beyond the borders of well-defined laboratories, thereby opening up towards exhibition spaces, political contexts, and public environments.

1. The Laboratory as Locus of Research

Laboratory studies, an important field within the science studies emphasize that we must understand scientific research as a situated practice. This implies a better understanding of how the particular material and structural, but also cultural and epistemic, qualities of a laboratory act upon and shape the possibilities and limitations of the research that can be conducted in such a laboratory: "*... laboratories are collective units that encapsulate within them a traffic of substances, materials, equipments, and observations ...*" (Knorr Cetina 1999, 38). Here the proactive, constructive qualities of scientific research are emphasized, arguing that "*... experiments deploy and implement a technology of intervention ...*" (Knorr Cetina, 1999, 37). Laboratories are the spaces where the artifacts, images, interfaces, methods, theories, materials, and processes are bundled and related, making it possible to perform specific experiments in the light of specific theories, answer unanswered research questions, and ask new research questions. This implies that "... *organizationally, experiments conduct 'science', while laboratories provide the (infra-)structure for carrying it out ... we need to conceive of laboratories as processes through which reconfigurations are negotiated, implemented, superseded, and replaced* ..." (Knorr Cetina 1999, 42 ff.). In the laboratory context, phenomena and their interpretation are iteratively transformed. For design research, this perspective indicates the importance of *defining in which "laboratories" and other creative "spaces" research takes place, based on the identification of the questions and answers, objects and images, and technologies and methods needed to enable the research process.*

2. Experimentation and Experimental Systems in Research

The science studies argue that experimentation has become the predominant method used in scientific research to act upon and shape reality in ways that make possible the exploration of research questions and the realization of specific research agendas. Experimentation is embedded in the creation and construction of specific systems, as specific assemblages of technologies, artifacts, representation tools and images, methods, research questions, disciplinary perspectives, ... , which together form an experimental system (Rheinberger 2001): "... *experimental systems are to be seen as the smallest integral working units of research. As such, they are systems of manipulation designed to give unknown answers to questions that the experimenters themselves are not yet able clearly to ask. Such setups are 'machines for making the future'* ..." (Rheinberger 2001, 28). Contrary to a perspective that sees experimentation itself as a clear indication of the inherently scientific nature of design practice and design research, it must argued that it is *through the construction of an experimental system, which allows the exploration of relevant research questions and the investigation of a research agenda, that design research has the potential to develop knowledge that will potentially qualify as scientific knowledge* (Grand 2010). In this way, experimental systems in scientific research and design research have a close relationship to central qualities of design practice, in that they make possible the intentional creation and materialization of research. As ways of exploring unanswered questions, experimental systems require tools and machines that allow the creation of the various experiments as they are intended to take place, as well as related imaging technologies, representation tools, and documentation strategies.

3. Collective Experiments and their Protocols

Contrary to the traditional view of experimental systems as well-defined laboratory spaces with clear boundaries, in recent years the science studies argue that experimental systems often do not remain within the boundaries of the laboratory (Nowotny 2008; Latour 1999; Felt, Nowotny & Taschwer 1995): today, many important societal issues and open questions are somewhat related to what could be called collective experiments (Latour 1999; 2001). Our societies are characterized by basic debates and controversial experiments concerning their possible futures, implying a high degree of complexity, due to the multiple perspectives, interests, concerns, issues, and approaches represented by the multiple parties involved. This implies "*That we are all engaged into a set of collective experiments that have spilled over the strict confines of the laboratories does not need more proof than the reading of the newspapers or the watching of the night TV news . . .* " (Latour 2001, 1). Our world can be seen as being engaged in multiple complex and collective experiments, concerning global warming, financial crises, genetic engineering, and ubiquitous computing. This view extends our understanding of what having a "knowledge society" means (Stehr 1994; Latour 1999), as well as sheds new light on the potential relevance of the so-called creative class (Florida 2002). Design research and design practice are already engaged in multiple collective experiments, from urbanism to interaction design. However, it is essential for design research to explicitly reflect and proactively contribute to the creation and *construction of such collective experiments, by establishing experimental systems that enable us to ask the right questions and to find robust answers, to map key controversies, and to visualize insightful patterns.*

The opportunities for rethinking the role of design practice in scientific research, as well as for redefining scientific research from the perspective of design, are thus not only important within the boundaries of traditional laboratory settings, but actually become a way of understanding, describing, structuring, and creating the experimental systems that our societies need in order to deal with their most controversial, essential, and complex questions and challenges. This is one of several major arguments we have introduced in favor of understanding research as design. Designerly ways of knowing are fundamental to any attempt to build experimental systems, and to playing a proactive role in those controversies and collective experiments. *Design practice and design research can be seen as creating and developing particular practices, tools, and methods of imagination, materialization, visualization, representation, and interaction relevant to those debates, while at the same time developing multiple new tools and methods for collectively dealing with possible futures in a complex world.*

Messiness and Order in Scientific Research

We are convinced that scientific research and design research must be seen as distributed, uncertain, controversial, multiple processes. This implies that any design research project must deal with these challenges; we argue that *an essential aspect of design research is that it deals with these qualities in a "designerly" way.* We have explored above what this means (see also Jonas 2001, in this book): "*. . . to a first approximation, then, science is an activity that involves the simultaneous orchestration of a wide range of appropriate literary and material arrangements . . .*" (Law 2004, 29). This also implies that "*. . . more generally, it is a set of arrangements for labeling, naming, and counting. It is a set of arrangements for getting from non-trace-like to trace-like form. It is a set of practices for shifting material modalities . . .*" (Law 2004, 29). As a consequence "*. . . methodological procedures and meticulous note-keeping are necessary . . . [researchers] are fixated on the business of keeping tabs on things. And if this fails, then the* *work of the laboratory fails . . .*" (Law 2004, 29). For design research, it is thus important to reflect on how particular ways of materializing and visualizing, and representing and projecting act on and shape the research process, as well as the potential outcomes of a design process. Furthermore, it is important to note that design research, and scientific research in general, are never just about dealing with the world as it is: "*. . . reality is neither independent nor anterior to its apparatus of production. . . . Realities are made. They are effects of the apparatuses of inscription . . .*" (Law 2004, 32).

However, this does not mean that "anything goes": "*. . . At the same time, since there are such apparatuses already in place, we also live in and experience a real world filled with real and more or less stable objects . . .* " (Law 2004, 32). This means that we constantly refer to artifacts and images, theories and perspectives, methods and tools, as taken-for-granted (Latour 1999). This is a simple matter of the impossibility of questioning everything at the same time, as well as of the cost of creating and establishing everything from scratch, something that holds true not just for research processes but for creation processes in general: "*. . . the argument is that undermining the relations embedded in received statements is expensive. The set of statements considered too costly to modify constitute what is referred to as reality . . .*" (Law 2004, 32). This implies that any project in design research and scientific research must decide what is to be taken for granted, and what is to be questioned; what is identified as real, and what is challenged: "*. . . these might range from things that everyone in question knows, through mundanities that no one notices until they stop happening, to matters and processes that are actively suppressed in order to produce the representations that are taken to report directly on realities . . .* " (Law 2004, 42). Design practice and design research are moving between what is taken for granted and what must be questioned, what is accepted and what is challenged, and what is identified as real and what is seen as a future possibility. In the last section of this essay, we take this as a starting point to identify a series of promising research strategies and future possibilities for design research.

Part 3: Design Fiction: Promising Strategies for Design Research

Design Fiction & Critical Design

To introduce our perspective on design research, we take two recent approaches in the area of design as a starting point: *Design Fiction* (Bleecker 2010) and *Critical Design* (Dunne & Raby 2001). As we will see, these two approaches bundle, assemble, relate, and integrate multiple qualities of design practice and scientific research, as we have discussed above. Furthermore, both approaches take a particular interest in referring to the particular qualities of designerly knowledge. They then open up new perspectives for design research and design practice by focusing on possible futures and potential perspectives for design research and design practice. Finally, they expand their relevance beyond the fields of design and design research by relating their research to technological innovation, scientific research, and experimentation in laboratories, as well as societal, cultural, and political issues and debates.

1. Design Fiction

"... *thinking through fiction to comprehend the action of design is a way to invigorate what design could be, beyond the routine, everyday notion of what design does* ..." (Bleecker 2010, 58). Design Fiction as an approach to design practice, as well as a perspective for design research, argues that it is a fundamental quality of design to rethink and reimagine what can be possible. This implies questioning assumptions about what the present and the future are for, what they contain, and what counts as an advancement "forward" toward a better, more habitable near-future world. Design Fiction thus uses design "... *to tell stories. It creates material artifacts that start conversations and suspend one's disbelief in what could be. It's a way of imagining a different kind of world by outlining the contours, rendering the artifacts as story props, using them to imagine new possibilities* ..." (Bleecker 2010, 62). For our discussion of research strategies in design research, it is interesting to realize that in the perspective of Design Fiction, "... *science and design and fact and fiction collapse together. They all wonder what a world would be like, if* ...". Given the generative qualities of experimental systems in scientific research, the importance of creation processes in design practice, and the complex interdependence between assuming reality and reflecting on the construction of realities, it becomes possible to identify various ways in which these different qualities relate. In simple terms, this leads to one central question, which underlies good design research and design practice: "... *a designer's motto should always be 'What if?'* ...", a comment made on "Nonobject", another recent research program, referring explicitly to Design Fiction (Lukic & Katz 2010).

2. Critical Design

In a related but distinct approach, Critical Design uses speculative design proposals to challenge narrow premises, assumptions, preconceptions, and givens about the role that products and technologies, designs and theories, and methods and artifacts play in everyday life (Dunne & Raby 2001; Dunne 1999). Therefore, design must be seen more as an attitude than anything else "... *to make us think, but also raising awareness, exposing assumptions, provoking action, sparking debate, even entertaining in an intellectual sort of way* ..." (Dunne & Raby, Critical Design FAQ). Referring to Manzini, an important reference in the context of our discussion, Critical Design argues that design is about visualizing "... *alternative future scenarios in ways that can be presented to the public, thus enabling democratic choices between the futures people actually want. Designers could then set about achieving these futures by developing new design strategies to direct industry to work with society* ... " (see Dunne 1999, xviii). More specifically, Critical Design deduces and sets out a series of qualities of design research, argued in the context of electronic artifacts, but with relevance for design practice and design research more generally: "... *One result of this research is a toolbox of concepts and ideas* ... *the most important elements of this approach are: going beyond optimization to explore critical and aesthetic roles for electronic products; using estrangement to open [up] the space between people and electronic products to discussion and criticism; designing alternative functions to draw attention to legal, cultural, and social rules; exploiting the unique narrative possibilities offered by electronic products; raising awareness of the electromagnetic qualities of our environment; and developing forms of engagement that avoid being didactic and utopian ...* " (Dunne 1999, 147).

The Design Fiction Method Toolbox

Based on our discussion of Design Fiction and Critical Design, and our perspectives on design as practice (see part 1), as well as on research as design (see part 3), we are able to deduce six dimensions and qualities of what we call the "Design Fiction Method Toolbox", a set of research strategies and design methods that allow us to translate our basic considerations into the practice of design research.

1. Project(ion)s: Design Fiction requires research methods, practices, and tools that make possible the *creation and construction of possible future worlds*, in relation to the actual world.

2. Materializations: Design Fiction searches for methods, practices, and tools that make it possible to *materialize those possible future worlds* in terms of images, artifacts, interfaces, and usages realized in diverse media.

3. Multitude: The Design Fiction Method Toolbox is characterized by an interest in a multitude of perspectives that go beyond one-sided ideological premises and make it possible to *map a multitude of possible perspectives.*

4. Processes: Design Fiction research methods and tools must be able to *represent, visualize, and document research and design as processes*, in order to trace and ground insights into what takes place in design research as process.

5. Systems: Design Fiction insists on the importance of understanding *experimentation processes as being generated through experimental systems*, making possible series of experiments, and exploring series of related hypotheses.

6. Reflexivity: The multiple representations in the context of the Design Fiction Method Toolbox *continuously challenge and advance the design research practices* themselves, in order to create new strategies, methods, and practices that fulfill these dimensions.

1. Project(ion)s: Creating and constructing possible future worlds

As discussed, *creating possible future worlds implies simultaneously a focus on the future and relating the future to the present and the past*: Critical Design (Dunne & Raby 2001; Dunne 1999), for example, is a promising approach, which is building artifacts that materialize *and visualize the often invisible dimensions* of new technologies (including, for example, electromagnetic fields), while *criticizing the existing technologies* and the ways they hide important features. Other design research practices are characterized by a focus on exploring specific research questions, in particular on *asking unanswerable questions*, as MVRDV does in their five-minute city project, where an unanswerable question triggers the development of unconventional and creative approaches for dealing with urbanistic themes (Maas 2003). Reinterpreting the present and past is another way of opening up new future possibilities, by transforming and translating *what is* into *what could be*. Fashion design is characterized by the continuous reinterpretation of existing collections as part of the creation of new collections (Shimizu & NHK 2005). This implies that not only our views into the future, but also our views and interpretations of the present, and even of the past, can be highly creative and imaginative.

2. Materializations: Materializing possible future worlds

Throughout our essay, we have emphasized that design practice and "designerly" ways of knowing are characterized by an attempt *to materialize central features, to make possible futures tangible and visible, to translate ideas and concepts into prototypes and sketches*, and to explore and understand new opportunities by building and constructing them. *Sketching* is the central approach in design, which advances at the interfaces between the future and the present, the possible and the actual, the imaginative and the real (Gänshirt 2007; Flusser 1996; 2008). Thus we learn from an interdisciplinary view of the multiple design practices that sketching takes many different forms: *drawing* on paper is the prototypical example, but in addition, building *simple models* in architectural design and industrial design follows a very similar direction, as does the *development of a mood board* in fashion design, or *simulating interactions* in the new media environment. Further, we can argue that the *ethnographic observation of design in use* is another way of exploring potentially inspiring new ways of realizing, visualizing, and embodying the future (Kelley 2001). In this perspective, the future is actually seen as always already taking place in the everyday activities of people using and misusing design for their own purposes; it is embedded in their quotidian practices.

3. Multitude: Mapping potentially relevant perspectives

It is essential to complement the first two strategies with an emphasis on the multiplicity and heterogeneity of possible futures, as well as on the *multitude of possible approaches and strategies in inventing, sketching, realizing, and visualizing those possible futures*. In this respect, we share the intuition of the sociology of translation (Latour 1999), which studies the underlying practices and processes of challenging the taken-for-granted, unquestioned, self-evident "nature" of the world as it is, while at the same time emphasizing and mapping the multitude of possible alternative worldviews. Interestingly, it is in this context that we see a central difference between the more ideological approaches to design research, which insist on the superiority of a particular approach or perspective (see the introduction, above), and Design Fiction, which emphasizes the importance of entering into controversies between different perspectives, methods and strategies as an important aspect of a research approach to design research. Thus the science studies argue that the systematic development of tools and methods for mapping multiple perspectives and related controversies are very important, but currently leave them underexplored (Latour & Weibel 2005). The recent interest in artistic and design practices in relation to scientific research indicates a growing interest in the advancement of scientific competencies and methods.

4. Processes: Representing research and design as processes

Interestingly, mapping controversies and representing perspectives is a precondition for advancement with respect to another fundamental area of design research: to advance in our understanding, description, explanation, and expertise in design research to the point of inventing and materializing, imagining and visualizing, creating and embodying new possible futures, we need *tools and methods that document and represent, map and visualize the research and design processes themselves* (Latour 2001). In many design fields, this emphasis on the design and research process is coming to the forefront of discussion: in urbanistic and architectural contexts we find *public debate based on models and computer simulations* (Franck & Franck 2008); in fashion design, exhibiting the *materiality, processuality, and multiplicity of design as a practice*, instead of overemphasizing outcomes (Maison Martin Margiela 2008); in iconic research, emphasizing the importance of integrating *sketches, drawings, models, and simulations* as crucial to our understanding of the resulting picture or installation (Boehm 2007). Further, the growing interest in exploring the potential of *programming design and experimentation processes*, as an important way of underscoring the processuality of design, is interesting to observe (Maeda 2000).

5. Systems: Experimentation generated through experimental systems

With a focus on programming, the inherently systemic nature of creation and experimentation is considered as a matter of course. As discussed above, design research can only benefit from the recent insights in the science studies if the *processual and systemic nature of experimentation and the importance of creating and establishing experimental systems are really understood* (Knorr Cetina 1999; Rheinberger 2001). In parallel, it becomes obvious that the design practices that are particularly important for advancing and conceptualizing design research are those which inherently consider experimental systems as their way of organizing practice and research: the current interest in *programming design processes* is obviously one way of advancing this research field (Maeda 2000). In parallel, *artistic processes exploring seriality* are important, as they help us to better understand the close interplay between shifting research questions and their relationships to shifting experimental arrangements (Calle 2003). Furthermore, we see a growing interest in *understanding archives as laboratories*, interpreted as realized and materialized series of artistic and designerly practices and processes (Bismarck 2002). Overall, we observe a growing interest in exploring design practice and design research as taking place in *laboratories*, which implies a specific materialization, a specific constellation of artifacts and technologies, methods and apparatuses, perspectives and visualizations. Only such systems make allowance for processuality and ensure the systematic representation of what is going on in the experimentation process (Obrist & Vanderlinden 1999).

6. Reflexivity: Challenging and advancing design research practices

It is essential to end this discussion of the Design Fiction Method Toolbox with a disclaimer: as soon as the different strategies and approaches to design research discussed here become taken-for-granted, self-evident, unquestioned strategies, they lose their "designerly" quality, which implies their focus on the creation, intention, projection, and invention of possible futures, as formulated in the first strategy (Simon 1996). We thus come full circle, back to our paraphrase of Latour—three things are particularly problematic in most discussions of design research: the word "design", the word "research", and the two words in combination—meaning that it is particularly problematic to assume that these three words are well defined and stable and can thus be taken for granted. Instead, we suggest that it is *crucial to constantly reinvent and imagine what the future of design research could be*, with implications for our views on design research as it takes place today (see Jonas 2011, in this book), as well as on our interpretations of the history of design research.

Epilogue: Possible Futures for Design Research

In this essay, we see the distinct contribution of design as a field of practice and research in *focusing on the world as it could be: what if?* However, while this perspective is important in many classical approaches to design research (Simon 1996; Bonsiepe 2007), as well as in multiple approaches to design practice (Lukic 2010; Auger 2010), it has not really been explored as the actual starting point for conducting, positioning, reflecting on, and practicing design research. We argue that *Design Fiction* (Bleecker 2010) and *Critical Design* (Dunne 1999), as ways of approaching design practice and design research, in relation to scientific research and technological innovation, as well as to societal, cultural, and political controversies, allow advancement in this direction, by explicitly identifying and discussing a *Design Fiction Method Toolbox* for design research. Design Fiction can thus benefit from the science studies: design research and scientific research in general can be interpreted as constructive and creative practices (Knorr Cetina 1999), structured as experimental systems (Rheinberger 2001), or embedded in transdisciplinary mode 2 constellations and multiple controversies and trading zones (Galison 1997). Emphasizing the processual and systemic nature of experimentation and translation (Callon 1986), and the importance of developing tools, methods, techniques, and media for mapping, representing, visualizing, and inscribing those experimental processes (Knorr Cetina 1999), Design Fiction makes it possible to open up a new research field of design research, which simultaneously leverages the *unique qualities of design as a practice*, and incorporates the quality criteria for productive and creative experimentation in scientific research: *research as design* (Glanville 1988).

REFERENCES

Auger, James (2010): "Alternative Presents and Speculative Futures" in: Swiss Design Network (ed.): *Negotiating Futures-Design Fiction*, Basel: HGK Basel, 42–57.

Baur, Ruedi (2009): *Intégral: Antizipieren–Hinterfragen–Einschreiben–Irritieren–Orientieren–Übersetzen–Unterscheiden*, Basel: Lars Müller.

Bijker, Wiebe E., Hughes, Thomas P. & Pinch, Trevor (eds., 1989): *The Social Construction of Technological Systems*, Cambridge, Massachusetts: MIT Press.

Bismarck, Beatrice von (ed., 2002). *Interarchive*, Köln: Walther König.

Bleecker, Julian (2010): "Design Fiction: From Props to Prototypes" in: Swiss Design Network (ed.): *Negotiating Futures-Design Fiction*, Basel: HGK Basel, 58–67.

Bleecker, Julian (2009): *Design Fiction: A short essay on design, science, fact and fiction*, Near Future Laboratory.

Boehm, Gottfried (2007): *Wie Bilder Sinn erzeugen: Die Macht des Zeigens*, Berlin: Berlin University Press.

Boland, Richard J. & Collopy, Fred (eds., 2004): *Managing as Designing*, Stanford: Stanford Business Books.

Boltanski, Luc (2009): *De la critique*, Paris: Gallimard.

Bolz, Norbert (2006): *Bang Design: Design-Manifest des 21. Jahrhunderts*, Hamburg: Trendbüro.

Bonsiepe, Gui (2007): "The Uneasy Relationship between Design and Design Research" in: Michel, Ralf (ed.): *Design Research Now*, Basel: Birkhäuser Verlag, 25–40.

Brooks, Rodney (2002): *Flesh and Machines*, New York: Pantheon Books.

Brown, Tim (2005): "Strategy by Design", *The Fast Company*, June.

Brown, Tim (2008): "Design Thinking", *Harvard Business Review*, June.

Brown, Tim (2009): *Change by Design*, New York: Harper Business.

Calle, Sophie (2003): *M'as tu vue*, Paris: Centre Pompidou.

Callon, Michel (1999): "Some Elements of a Sociology of Translation" in: Biagioli, Mario (ed.): *The Science Studies Reader*, New York: Routledge.

Cross, Nigel (2006): *Designerly Ways of Knowing*, New York: Springer.

Dunne, Anthony & Raby, Fiona (2001): *Design Noir: The Secret Life of Electronic Objects*, Basel: Birkhäuser Verlag.

Dunne, Anthony & Raby, Fiona: *Critical Design FAQ*.

Dunne, Anthony (1999): *Hertzian Tales: Electronic products, aesthetic experience, and critical design*: Boston, Massachusetts: MIT Press.

Elkana, Yehuda (1986): *Anthropologie der Erkenntnis*, Frankfurt am Main: Suhrkamp.

Felt, Ulrike, Nowotny, Helga & Taschwer, Klaus (1995): *Wissenschaftsforschung: Eine Einführung*, Frankfurt am Main: Campus

Findeli, Alain (2007): "Die projektgeleitete Forschung: Eine Methode der Designforschung" in: Michel, Ralf (ed.). *Erstes Design Forschungssymposium. Basel*, Zürich: Swiss Design Network, 40–51.

Florida, Richard (2002): *The Rise of the Creative Class*. New York: Basic Books.

Flusser, Vilem (1996): *Vom Stand der Dinge: Eine kleine Philosophie des Design*, Göttigen: Steidl Verlag.

Flusser, Vilem (2008): *Kommunikologie weiter denken: Die Bochumer Vorlesungen*, Franfurt am Main: Fischer Taschenbuch Verlag.

Foucault, Michel (1971): *L'ordre du discours*, Paris: Gallimard.

Franck, Georg & Franck, Dorothea (2008): *Architektonische Qualität*, München: Carl Hanser Verlag.

Galison, Peter (1997): "Trading Zone. Coordinating Action and Belief" in: Biagioli, Mario (ed.): *The Science Studies Reader*, New York: Routledge.

Gänshirt, Christian (2007): *Werkzeuge für Ideen: Einführung ins Architektonische Entwerfen*, Basel: Birkhäuser Verlag.

Gehlen, Arnold (1961): *Anthropologische Forschung*, Hamburg: Rowolt.
Gehlen, Arnold (2004): *Urmensch und Spätkultur: Philosophische Ergebnisse und Aussagen*, Frankfurt am Main: Klostermann.
Gibbons, Michael, Limoges, Camille, Nowotny, Helga, Schwartzman, Simon, Scott, Peter & Trow, Martin (1994): *The new production of knowledge*, London: Sage Publications.
Glanville, Ranulf (1988): *Objekte*, Berlin: Merve Verlag.
Grand, Simon (2009): "Design Fiction und unternehmerische Strategien" in: Martin Wiedmer (ed.): *Design Fiction: Perspektiven für Forschung in Kunst und Design*, Basel: HGK Basel, 21–26.
Grand, Simon (2010): "Strategy Design: Design Practices for Entrepreneurial Strategizing" in: Samiyeh, Michael (ed.): *Creating Desired Futures: How Design Thinking Innovates Business*, Basel: Birkhäuser Verlag.
Hernes, Tor (2008): *Understanding Organization as Process: Theory for a Tangled World*, New York: Routledge.
Honneth, Axel & Joas, Hans (1980): *Soziales Handeln und menschliche Natur*, Frankfurt am Main: Campus.
Jonas, Wolfgang (2001): *Entwerfen als "sumpfiger Grund" unseres Konzepts von Menschen und Natur (-Wissenschaften)*, Köln: Salon Verlag, 114–125.
Jonas, Wolfgang (2007): "Design Research and its Meaning to the Methodological Development of the Discipline" in: Michel, Ralf (ed.): *Design Research Now*, Basel: Birkhäuser Verlag, 187–206.
Jonas, Wolfgang (2011): in this volume.
Kelley, Tom (2001): *The Art of Innovation*, New York: Random House.
Knorr Cetina, Karin (1999): *Epistemic Cultures: How the Sciences Make Knowledge*, Cambridge, Massachusetts: Harvard University Press.
Krippendorff, Klaus (2007): "Design Research: An Oxymoron?" in: Michel, Ralf (ed.): *Design Research Now*, Basel: Birkhäuser Verlag, 67–80.
Latour, Bruno & Weibel, Peter (2002): *Making Things Public: Atmospheres of Democracy*, Cambridge, Massachusetts: MIT Press.
Latour, Bruno (1999): *Pandora's Hope*, Cambridge, Massachusetts: Harvard University Press.
Latour, Bruno (2001): *Iconoclash: Gibt es eine Welt jenseits des Bilderkrieges?* Berlin: Merve Verlag.
Latour, Bruno (2004): Von ‚Tatsachen' zu ‚Sachverhalten': Wie sollen die neuen kollektiven Experimente protokolliert werden?" in Schmidgen, Henning, Geimer, Peter & Dierig, Sven (eds.): *Kultur im Experiment*, Berlin: Kulturverlag Kadmos.
Laurel, Brenda (ed., 2003): *Design Research: Methods and Perspectives*, Cambridge, Massachusetts: MIT Press.
Law, John (2004): *After Method: Mess in Social Science Research*, London: Routledge.
Lukic, Branco & Katz, Barry, M. (2010): *Nonobject*, Cambridge, Massachusetts: MIT Press.
Maar, Christa & Burda, Hubert (eds., 2004): *Iconic Turn: Die neue Macht der Bilder*, Köln: Dumont.
Maas, Winy (2003): *The Five Minute City*, Rotterdam: Episode Publishers.
Maeda, John (2000): *Maeda @ Maeda*, New York: Universe Publishing.
Maison Martin Margiela (2008): 20: *The Exhibition*, Antwerp: MOMU.
Nowotny, Helga & Testa, Giuseppe (2009): *Die gläsernen Gene. Die Erfindung des Individuums im molekularen Zeitalter*, Frankfurt am Main: Suhrkamp.
Nowotny, Helga (2008): *Unersättliche Neugier: Innovation in einer fragile Zukunft*, Berlin: Kadmos.
Obrist, Hans-Ulrich & Vanderlinden, Barbara (1999): *Laboratorium*, Antwerpen: Du Mont.

Orlikowski, Wanda (2000): "Using Technology and Constituting Structures: A Practice Lense for Studying Technology in Organizations", *Organization Science* 11, 4: 404–428.
Orlikowski, Wanda (2002): "Knowing in Practice: Enacting a Collective Capability in Distributed Organizing", *Organization Science*, 13 4: 249–273.
Pfeifer, Rolf & Bongard, Josh (2007): *How the body shapes the way we think*, Cambridge, Massachusetts: MIT Press.
Rheinberger, Hans-Jörg (1997): *Toward a History of Epistemic Things. Synthesizing Proteins in the Test Tube*, Stanford: Stanford University Press.
Schatzki, Theodore R., Knorr Cetina, Karin & von Savigny, Eike (eds., 2001): *The Practice Turn in Contemporary Theory*, London: Routledge.
Schumpeter, Joseph (1942): *Capitalism, Socialism, and Freedom*, New York: Harper & Brothers.
Shimizu, Sanae & NHK (2005): *Unlimited: Comme des Garçons*, Tokyo: Heibonsha.
Simon, Herbert A. (1996): *The Sciences of the Artificial*, Cambridge, Massachusetts: MIT Press.
Sloterdijk, Peter (2006): "Das Zeug zur Macht" in: Seltmann, Gerhard & Lippert, Werner (eds.): *Entry Paradise: Neue Welten des Designs*, Basel: Birkhäuser Verlag.
Sloterdijk, Peter (2009): *Scheintod im Denken: Von Philosophie und Wissenschaft als Übung*, Frankfurt am Main: Suhrkamp.
Stehr, Nico (1994): *Knowledge Society*, London: Sage.
Tsoukas, Haridimos (1991): "The Missing Link: A Transformational View of Metaphors in Organizational Science", *Academy of Management Review* 16(3): 566–585.
Utterback, James, Vedin, Bengt-Arne, Alvarez, Eduardo, Ekman, Sten, Sanderson, Susan Walsh, Tether, Bruce & Verganti, Roberto (2006): *Design-Inspired Innovation*, Singapore: World Scientific.
Verganti, Roberto (2009): *Design Driven Innovation*, Boston, Massachusetts: Harvard Business Press.
von Hippel, Eric (2005): *Democratizing Innovation*, Cambridge, Massachusetts: MIT Press.

FRIEDRICH NIETZSCHE

THE JOYFUL WISDOM

Finally (that the most essential may not remain unsaid), one comes back out of such abysses, out of such severe sickness, and out of the sickness of strong suspicion—*new-born*, with the skin cast; more sensitive, more wicked, with a finer taste for joy, with a more delicate tongue for all good things, with a merrier disposition, with a second and more dangerous innocence in joy; more childish at the same time, and a hundred times more refined than ever before. Oh, how repugnant to us now is pleasure, coarse, dull, drab pleasure, as the pleasure-seekers, our "cultured" classes, our rich and ruling classes, usually understand it ! How malignantly we now listen to the great holiday-hubbub with which "cultured people" and city-men at present allow themselves to be forced to "spiritual enjoyment" by art, books, and music, with the help of spirituous liquors! How the theatrical cry of passion now pains our ear, how strange to our taste has all the romantic riot and sensuous bustle which the cultured populace love become (together with their aspirations after the exalted, the elevated, and the intricate)! No, if we convalescents need an art at all, it is *another* art—a mocking, light, volatile, divinely serene, divinely ingenious art, which blazes up like a clear flame, into a cloudless heaven ! Above all, an art for artists, only for artists ! We at last know better what is first of all necessary *for it*—namely, cheerfulness, *every* kind of cheerfulness, my friends ! also as artists :—I should like to prove it. We now know something too well, we men of knowledge : oh, how well we are now learning to forget and *not* know, as artists ! And as to our future, we are not likely to be found again in the tracks of those Egyptian youths who at night make the temples unsafe, embrace statues, and would fain unveil, uncover, and put in clear light, everything which for good reasons is kept concealed.[1] No, we have got disgusted with this bad taste, this will to truth, to "truth at all costs," this youthful madness in the love of truth : we are now too experienced, too serious, too joyful, too singed, too profound for that. . . .We no longer believe that truth remains truth when the veil is withdrawn from it : we have lived long enough to believe this. At present we regard it as a matter of propriety not to be anxious either to see everything naked, or to be present at everything, or to understand and "know" everything. "Is it true that the good God is everywhere present ?" asked a little girl of her mother : "I think that is indecent" :—a hint to philosophers ! One should have more reverence for the *shatne-facedness* with which nature has concealed herself behind enigmas and motley uncertainties. Perhaps truth is a woman who has reasons for not showing her reasons ? Perhaps her name is Baubo, to speak in Greek ? . . . Oh, those Greeks ! They knew how *to live :* for that purpose it is necessary to keep bravely to the surface, the fold and the skin ; to worship appearance, to believe in forms, tones,

[1] An allusion to Schiller's poem: "The Veiled Image of Sais." – Tr.

and words, in the whole Olympus of appearance ! Those Greeks were superficial—*from profundity!* And are we not coming back precisely to this point, we dare-devils of the spirit, who have scaled the highest and most dangerous peak of contemporary thought, and have looked around us from it, have *looked down* from it ? Are we not precisely in this respect—Greeks? Worshippers of forms, of tones, and of words ? And precisely on that account-artists ?

Ruta, near Genoa
Autumn, 1886.

from: Nietzsche, Friedrich (1887), The Joyful Wisdom

PETER GALISON

1 TRADING ZONE

Coordinating Action and Belief

Part I: Intercalation

Introduction: The Many Cultures of Physics

I will argue this: science is disunified, and—against our first intuitions—it is precisely the *dis*unification of science that underpins its strength and stability. This argument stands in opposition to the tenets of two well-established philosophical movements: the logical positivists of the 1920s and 1930s who argued that unification underlies the coherence and stability of the sciences, and the antipositiviste of the 1950s and 1960s who contended that disunification implies instability. In *Image and Logic,* I have tried to bring out just how partial a theory-centered, single culture view of physics must be. Forms of work, modes of demonstration, ontologicai commitment—all differ among the many traditions that compose physics at any given time in the twentieth century. In this chapter, drawing on related work in the history and philosophy of science, I will argue that even specialties within physics cannot be considered as homogeneous communities. Returning to the idea of intuition I have sketched elsewhere, I want to reflect at greater length on a description of physics that would neither be unified nor splintered into isolated fragments. I will call this multicultural history of the development of physics *intercalated,* because the many traditions coordinate with one another without homogenization. Different finite traditions of theorizing, experimenting, instrument making, and engineering meet—even transform one another—but for all that they do not lose their separate identities and practices. [...]

These considerations so exacerbated the problem that it seemed as if any two cultures (groups with very different systems of symbols, and procedures for their manipulation) would seem utterly condemned to passing one another without any possibility of significant interaction. But here we can learn from the anthropologists who regularly study unlike cultures that do interna, most notably by trade. Two groups can agree on rules of exchange even if they ascribe utterly different significance to the objects being exchanged; they may even disagree on the meaning of the exchange process itself. Nonetheless, the trading partners can hammer out a *local* coordination despite vast *global* differences. In an even more sophisticated way, cultures in interaction frequently establish contact languages, systems of discourse that can vary from the most function-specific jargons through semi-specific pidgins, to full-fledged Creoles rich enough to support activities as complex as poetry and metalinguistic reflection. The anthropological picture is relevant here. For in focusing on local coordination, not global meaning, I think one can understand the way engineers, experimenters, and theorists interact. At last I come to the connection between place, exchange, and knowledge production. But instead of looking at laboratories simply as the place where experimental information and strategies are generated, my concern is with the site—partly symbolic and partly spatial—where the local coordination between beliefs and action takes place. It is a domain I will call the trading zone. [...]

Each subculture has its own rhythms of change, each has its own standards of demonstration, and each is embedded differently in the wider culture of institutions, practices, inventions, and ideas.[1]

Thus for historical reasons, instead of searching for a positivist central metaphor grounded in observation, or an antipositivist central metaphor grounded in theory, I suggest that we admit a wider class of periodization schemes, in which the three levels are *intercalated* (see Figure 3).

Different quasi-autonomous traditions carry their own periodizations. There are four facets of this open-ended model that merit attention. First, it is tripartite, granting (or at least offering the possibility of granting) a partial autonomy to instrumentation, experimentation, and theory. It is contingent, not preordained, that each subculture be represented separately as one can easily identify moments in the history of physics where the instrument makers and the experimentalists (to give one example) were not truly distinct. Nor is it *always* the case that break points occur separately. And there are many times when there were competing experimental subcultures each working in the same domain (bubble chamber users and spark chamber users, for example). Second, this *class* of central metaphors incorporates one of the key insights of the antipositivists: *there is no absolutely continuous basis in observation.* Both the level of experimentation and the level of instrumentation have their break points, just as theory does. Third, the local continuities are *intercalated*—we do not expect to see the abrupt changes of theory, experimentation, and instrumentation to occur simultaneously; in any case it is a matter of historical investigation to determine if they (contingently) do line up. Indeed, there are good reasons to expect that at the moment one stratum splits, workers in the others will do what they can to deploy accepted procedures that allow them to study the split before and after—when a radically new theory is introduced, we would expect experimenters to deploy their best-established instruments,

	instrument_1	instrument_2	instrument_3
theory_1	theory_2	theory_3	
	experiment_1	experiment_2	experiment_3
time →			

Figure 3
Intercalated Periodization

not their unproven ones. Fourth, we expect a rough *parity* among the strata—no one level is privileged, no one subculture has the special position of narrating the right development of the field or serving as the reduction basis (the intercalated strands should really be drawn in three dimensions so no one is on top and each borders on the other two). Just as a bricklayer would not stack set the bricks for fear his whole building would collapse, each individual (or research group) does what it can to set breaks in one practice cluster against continuities in others. As a result of such local actions (not by global planning), the community as a whole does not stack periodize its subcultures. [...]

Part II. The Trading Zone

The Locality of Exchange

In an effort to capture both the differences between the subcultures and the felt possibility of communication, consider again the picture of intercalated periodizations discussed earlier but now focus on the boundaries between the strata. To characterize the interaction between the subcultures of instrumentation, experiment, and theory, I want to pursue the idea that these really are subcultures of the larger culture of physics. Like two cultures, distinct but living near enough to trade, they can share some activities while diverging on many others. In particular, the two cultures may bring to what I will call the *trading zone* objects that carry radically different significance for the donor and recipient. What is crucial is that in the highly local context of the trading zone, *despite* the differences in classification, significance, and standards of demonstration, the two groups can collaborate. They can come to a consensus about the procedure of exchange, about the mechanisms to determine when the goods are "equal" to one another. They can even both understand that the continuation of exchange is a prerequisite to the survival of the larger culture of which they are part.

I intend the term trading zone to be taken seriously, as a social and intellectual mortar binding together the disunified traditions of experimenting, theorizing, and instrument building. Anthropologists are familiar with different cultures encountering one another through trade, even when the significance of the objects traded—and of the trade itself—may be utterly different for the two sides. For example, in the southern Cauca Valley, in Colombia, the mostly black peasants, descended from slaves, maintain a rich culture permeated with magical cycles, sorcery, and curing. They are also in constant contact with the powerful forces of the landowning classes: some of the peasants run shops, others work on the vast sugarcane farms. Daily life includes many levels of exchange between the two sides, in the purchase of goods, the payment of rent, and the disbursement of wages. And within this trading zone both sides are perfectly capable of working within established behavioral patterns. But the *understanding* each side has of the exchange of money is utterly different. For the white landowners, money is "neutral" and has a variety of natural properties; for example, it can accumulate into capital—money begets money. For the black peasants, funds obtained in certain ways have animistic, moral properties, though perhaps none more striking than the practice of the secret baptism of money. In this ritual, a godparent-to-be hides a peso note in his or her hand, while the Catholic priest baptizes the infant. According to local belief, the peso bill—rather than the child—is consequently baptized, the bill acquires the child's name, and the godparent-to-be becomes the godparent of the bill. While putting the bill into circulation, the owner quietly calls it by its name three times and the faithful pesos will return to the owner, accompanied by their kin, usually from the pocket of the recipient. So, when we narrow our gaze to the peasant buying eggs in a landowner's shop we may see two people, perfectly harmoniously exchanging items. In fact, they depend on the exchange for survival. Out of our narrow view, however, are two vastly different symbolic and cultural systems, embedding two perfectly incompatible valuations and understandings of the objects exchanged.[2]

In our case, theorists trade experimental predictions for experimentalists' results. Two things are noteworthy about the exchange. First, the two subcultures may altogether disagree about the implications of the information exchanged or its epistemic status. For example, as we have seen, theorists may predict the existence of an entity with profound conviction because it is inextricably tied to central tenets of their practice—for example, group symmetry, naturalness, renormalizability, covariance, or unitarity. The experimentalist may receive the prediction as something quite different, perhaps as no more than another curious hypothesis to try out on the next run of the data-analysis program. But despite these sharp differences, it is striking that there is a context *within* which there is a great deal of consensus. In this trading zone, phenomena are discussed by both sides. It is here that we find the classic encounters of experiment with theory: particle decays, fission, fusion, pulsars, magnetostriction, the creep effect, second sound, lasing, magnetic deflection, and so on. It is the existence of such trading zones, and the highly constrained negotiations that proceed within them, that bind the otherwise disparate subcultures together. [. . .]

Let me conclude with a metaphor. For years physicists and engineers harbored a profound mistrust of disorder. They searched for reliability in crystals rather than disordered materials, and strength in pure substances rather than laminated ones. Suddenly, in the last few years, in a quiet upheaval, they discovered that the classical vision had it backward: the electronic properties of crystals were fine until—*because* of their order—they failed catastrophically. It was amorphous semiconductors, with their *disordered* atoms, that gave the consistent responses needed for the modern era of electronics. Structural engineers were slow to learn the same lesson. The strongest materials were not pure—they were laminated; when they failed microscopically, they held in bulk. To a different end, in 1868 Charles Sanders Peirce invoked the image of a cable. I find his use evocative in just the right way: "Philosophy ought to imitate the successful sciences in its methods. . . . [T]o trust . . . rather to the multitude and variety of its arguments than to the conclusiveness of any one. Its reasoning should not form a chain which is no stronger than its weakest link, but a cable whose fibres may be ever so slender, provided they are sufficiently numerous and intimately connected."[3] With its intertwined strands, the cable gains its strength not by having a single, golden thread that winds its way through the whole. No one strand defines the whole. Rather, the great steel cables gripping the massive bridges of Peirce's time were made strong by the interleaving of many limited strands, no one of which held all the weight. Decades later, Wittgenstein used the same metaphor now cast in the image of thread, as he reflected on what it meant to have a concept. "We extend our concept of number as in spinning a thread we twist fibre on fibre. And the strength of the thread does not reside in the fact that some one fibre runs through its whole length, but in the overlapping of many fibres."[4] Concepts, practices, and arguments will not halt at the door of a conceptual scheme or its historical instantiation: they continue, piecewise.

These analogies cut deep. It is the *disorder* of the scientific community—the laminated, finite, partially independent strata supporting one another; it is the disunification of science—the intercalation of *different* patterns of argument—that is responsible for its strength and coherence. It is an intercalation that extends even further down—even within the stratum of instruments we have seen mimetic and analytic traditions as separate and then combining, image and logic compering then merging. So too could we see divisions within theory—confrontational views about symmetries, field theory, S-matrix theory, for example—as one incompletely overlapped the other.

But ultimately the cable metaphor too takes itself apart, for Peirce insists that the strands not only be "sufficiently numerous" but also "intimately connected." In the cable, that connection is mere physical adjacency, a relation unhelpful in explicating the ties that bind concepts, arguments, instruments, and scientific subcultures. No mechanical analogy will ever be sufficient to do that because it is by coordinating different symbolic and material actions that people create the binding culture of science. All metaphors come to an end.

NOTES

1 As we search to locate scientific activities in their context, it is extremely important to recognize that for many activities there is no single context. By viewing the collaboration among people in a laboratory not as a melding of identities but as a coordination among subcultures we can see the actors as separately embedded in their respective, wider worlds. E.g., in the huge bubble chamber laboratory at Berkeley one has to see the assembled workers as coming together from the AEC's secret world of nuclear weapons and from an arcane theoretical culture of university physics. The culture they partially construct at the junction is what I have in mind by the "trading zone."
2 The example of the secret baptism of money is from M. Taussig, *The Devil and Commodity Fetishism in South America* (Chapel Hill: University of North Carolina Press, 1980), ch. 7.
3 Charles Sanders Peirce, "Some Consequences of Four Incapacities," in *writings of Charles Sanders Peirce, A Chronological Edition, Vol. 2, 1867-1871* (Bloomington: Indiana University Press,1984), 213.
4 Ludwing Wittgenstein, *Philosophical Investigations,* 2d ed., trans. G. E. M. Anscombe (Oxford: Blackwell, 1958), par. 67.

Michael Gibbons, Camille Limoges, Helga Nowotny, Simon Schwartzman, Peter Scott, Martin Trow

2 THE NEW PRODUCTION OF KNOWLEDGE

The dynamics of science and research in contemporary societies

Introduction

This volume is devoted to exploring changes in the mode of knowledge production in contemporary society. Its scope is broad, concerned with the social sciences and the humanities as well as with science and technology, though fewer pages are given to the former than to the latter. A number of attributes have been identified which suggest that the way in which knowledge is being produced is beginning to change. To the extent that these attributes occur across a wide range of scientific and scholarly activity, and persist through time they may be said to constitute trends in the way knowledge is produced. No judgement is made as to the value of these trends—that is, whether they are good and to be encouraged, or bad and resisted—but it does appear that they occur most frequently in those areas which currently define the frontier and among those who are regarded as leaders in their various fields. Insofar as the evidence seems to say that most of the advances in science have been made by 5 per cent of the population of practising scientists, these trends, because they seem to involve the intellectual leaders, probably ought not to be ignored.

It is the thesis of this book that these trends do amount, not singly but in their interaction and combination, to a transformation in the mode of knowledge production. The nature of this transformation is elaborated for science, in Chapter 1; for technology in Chapter 2; in Chapter 4 for the humanities; and for the social sciences throughout the text. The transformation is described in terms of the emergence alongside traditional modes of knowledge production that we will call Mode 2. By contrast with traditional knowledge, which we will call Mode 1, generated within a disciplinary, primarily cognitive, context, Mode 2 knowledge is created in broader, transdisciplinary social and economic contexts. The aim of introducing the two modes is essentially heuristic in that they clarify the similarities and differences between the attributes of each and help us understand and explain trends that can be observed in all modem societies. The emergence of Mode 2, we believe, is profound and calls into question the adequacy of familiar knowledge producing institutions, whether universities, government research establishments, or corporate laboratories.

[...]

Some Attributes of Knowledge Production in Mode 2

Knowledge Produced in the Context of Application

The relevant contrast here is between problem solving which is carried out following the codes of practice relevant to a particular discipline and problem solving which is organised around a particular application. In the former, the context is defined in relation to the cognitive and social norms that govern basic research or academic science. Latterly, this has tended to imply knowledge production carried out in the absence of some practical goal. In Mode 2, by contrast, knowledge results from a broader range of considerations. Such knowledge is intended to be useful to someone whether in industry or government, or society more generally and this imperative is present from the beginning. Knowledge is always produced under an aspect of continuous negotiation and it will not be produced unless and until the interests of the various actors are included. Such is the context of application. Application, in this sense is not product development carried out for industry and the processes or markets that operate to determine what knowledge is produced are much broader than is normally implied when one speaks about taking ideas to the marketplace. None the less, knowledge production in Mode 2 is the outcome of a process in which supply and demand factors can be said to operate, but the sources of supply are increasingly diverse, as are the demands for differentiated forms of specialist knowledge. Such processes or markets specify what we mean by the context of application. Because they include much more than commercial considerations, it might be said that in Mode 2 science has gone beyond the market! Knowledge production becomes diffused throughout society. This is why we also speak of socially distributed knowledge.

Research carried out in the context of application might be said to characterise a number of disciplines in the applied sciences and engineering—for example, chemical engineering, aeronautical engineering or, more recently, computer science. Historically these sciences became established in universities but, strictly speaking, they cannot be called applied sciences, because it was precisely the lack of the relevant science that called them into being. They were genuinely new forms of knowledge though not necessarily of knowledge production because, they too, soon became the sites of disciplinary-based knowledge production in the style of Mode 1. These applied disciplines share with Mode 2 some aspects of the attribute of knowledge produced in the context of application. But, in Mode 2 the context is more complex. It is shaped by a more diverse set of intellectual and social demands than was the case in many applied sciences while it may give rise to genuine basic research.

Transdisciplinarity

Mode 2 does more than assemble a diverse range of specialists to work in teams on problems in a complex applications oriented environment. To qualify as a specific form of knowledge production it is essential that enquiry be guided by specifiable consensus as to appropriate cognitive and social practice. In Mode 2, the consensus is conditioned by the context of application and evolves with it. The determinants of a potential solution involve the integration of different skills in a framework of action but the consensus may be only temporary depending on how well it conforms to the requirements set by the specific context of application. In Mode 2, the shape of the final solution will normally be beyond that of any single contributing discipline. It will be transdisciplinarity.

Transdisciplinarity has four distinct features. First, it develops a distinct but evolving framework to guide problem solving efforts. This is generated and sustained in the context of application and not developed first and then applied to that context later by a different group of practitioners. The solution does not arise solely, or even mainly, from the application of knowledge that already exists. Although elements of existing knowledge must have entered into it, genuine creativity is involved and the theoretical consensus, once attained cannot easily be reduced to disciplinary parts.

Second, because the solution comprises both empirical and theoretical components it is undeniably a contribution to knowledge, though not necessarily disciplinary knowledge. Though it has emerged from a particular context of application, transdisciplinary knowledge develops its own distinct theoretical structures, research methods and modes of practice, though they may not be located on the prevailing disciplinary map. The effort is cumulative, though the direction of accumulation may travel in a number of different directions after a major problem has been solved.

Third, unlike Mode 1 where results are communicated through institutional channels, the results are communicated to those who have participated in the course of that participation and so, in a sense, the diffusion of the results is initially accomplished in the process of their production. Subsequent diffusion occurs primarily as the original practitioners move to new problem contexts rather than through reporting results in professional journals or at conferences. Even though problem contexts are transient, and problem solvers highly mobile, communication networks tend to persist and the knowledge contained in them is available to enter into further configurations.

Fourth, transdisciplinarity is dynamic. It is problem solving capability on the move. A particular solution can become the cognitive site from which further advances can be made, but where this knowledge will be used next and how it will develop are as difficult to predict as are the possible applications that might arise from discipline-based research. Mode 2 is marked especially but not exclusively by the ever closer interaction of knowledge production with a succession of problem contexts. As with discoveries in Mode 1 one discovery may build upon another but in Mode 2, the discoveries lie outside the confines of any particular discipline and practitioners need not return to it for validation. New knowledge produced in this way may not fit easily into any one of the disciplines that contributed to the solution. Nor may it be easily referred to particular disciplinary institutions or recorded as disciplinary contributions. In Mode 2, communications in ever new configurations are crucial. Communication links are maintained partly through formal and partly through informal channels.

Heterogeneity and Organisational Diversity

Mode 2 knowledge production is heterogeneous in terms of the skills and experience people bring to it. The composition of a problem solving team changes over time as requirements evolve. This is not planned or coordinated by any central body. As with Mode 1, challenging problems emerge, if not randomly, then in a way which makes their anticipation very difficult. Accordingly, it is marked by:

1. An increase in the number of potential sites where knowledge can be created; no longer only universities and colleges, but non-university institutes, research centres, government agencies, industrial laboratories, think-tanks, consultancies, in their interaction.

2. The linking together of sites in a variety of ways—electronically, organisationally, socially, informally—through functioning networks of communication.

3. The simultaneous differentiation, at these sites, of fields and areas of study into finer and finer specialities. The recombination and reconfiguration of these subfields form the bases for new forms of useful knowledge. Over time, knowledge production moves increasingly away from traditional disciplinary activity into new societal contexts.

In Mode 2, flexibility and response time are the crucial factors and because of this the types of organisations used to tackle these problems may vary greatly. New forms of organisation have emerged to accommodate the changing and transitory nature of the problems Mode 2 addresses. Characteristically, in Mode 2 research groups are less firmly institutionalised; people come together in temporary work teams and networks which dissolve when a problem is solved or redefined. Members may then reassemble in different groups involving different people, often in different loci, around different problems. The experience gathered in this process creates a competence which becomes highly valued and which is transferred to new contexts. Though problems may be transient and groups short-lived, the organisation and communication pattern persists as a matrix from which further groups and networks, dedicated to different problems, will be formed. Mode 2 knowledge is thus created in a great variety of organisations and institutions, including multinational firms, network firms, small hi-tech firms based on a particular technology, government, institutions, research universities, laboratories and institutes as well as national and international research programmes. In such environments the patterns of funding exhibit a similar diversity, being assembled from a variety of organisations with a diverse range of requirements and expectations which, in turn, enter into the context of application.

Social Accountability and Reflexivity

In recent years, growing public concern about issues to do with the environment, health, communications, privacy and procreation, and so forth, have had the effect of stimulating the growth of knowledge production in Mode 2. Growing awareness about the variety of ways in which advances in science and technology can affect the public interest has increased the number of groups that wish to influence the outcome of the research process. This is reflected in the varied composition of the research teams. Social scientists work alongside natural scientists, engineers, lawyers and businesspeople because the nature of the problems requires it. Social accountability permeates the whole knowledge production process. It is reflected not only in interpretation and diffusion of results but also in the definition of the problem and the setting of research priorities. An expanding number of interest, and so-called concerned, groups are demanding representation in the setting of the policy agenda as well as in the subsequent decision making process. In Mode 2, sensitivity to the impact of the research is built in from the start. It forms part of the context of application.

Contrary to what one might expect, working in the context of application increases the sensitivity of scientists and technologists to the broader implications of what they are doing. Operating in Mode 2 makes all participants more reflexive. This is because the issue on which research is based cannot be answered in scientific and technical terms alone. The research towards the resolution of these types of problem has to incorporate options for the implementation of the solutions and these are bound to touch the values and preferences of different individuals and groups that have been seen as traditionally outside of the scientific and technological system. They can now become active agents in the definition and solution of problems as well as in the evaluation of performance. This is expressed partly in terms of the need for greater social accountability, but it also means that the individuals themselves cannot function effectively without reflecting—trying to operate from the standpoint of—all the actors involved. The deepening of understanding that this brings, in turn has an effect on what is considered worthwhile doing and, hence, on the structure of the research itself. Reflection of the values implied in human aspirations and projects has been a traditional concern of the humanities. As reflexivity within the research process spreads, the humanities too are experiencing an increase in demand for the sorts of knowledge they have to offer.

Traditionally, this has been the function of the humanities, but over the years the supply side—departments of philosophy, anthropology, history—of such reflexivity has become disconnected from the demand side—that is from businesspeople, engineers, doctors, regulatory agencies, and the larger public who need practical or ethical guidance on a vast range of issues (for example, pressures on the traditional humanities for culturally sensitive scenarios, and on legal studies for an empirically grounded ethics, the construction of ethnic histories, and the analysis of gender issues).

Quality Control

Criteria to assess the quality of the work and the teams that carry out research in Mode 2 differ from those of more traditional, disciplinary science. Quality in Mode 1 is determined essentially through the peer review judgements about the contributions made by individuals. Control is maintained by careful selection of those judged competent to act as peers which is in part determined by their previous contributions to their discipline. So, the peer review process is one in which quality and control mutually re-enforce one another. It has both cognitive and social dimensions, in that there is professional control over what problems and techniques are deemed important to work on as well as who is qualified to pursue their solution. In disciplinary science, peer review operates to channel individuals to work on problems judged to be central to the advance of the discipline. These problems are defined largely in terms of criteria which reflect the intellectual interests and preoccupations of the discipline and its gatekeepers.

In Mode 2, additional criteria are added through the context of application which now incorporates a diverse range of intellectual interests as well as other social, economic, or political ones. To the criterion of intellectual interest and its interaction, further questions are posed, such as 'Will the solution, if found, be competitive in the market?' 'Will it be cost effective?', 'Will it be socially acceptable?' Quality is determined by a wider set of criteria which reflects the broadening social composition of the review system. This implies that 'good science' is more difficult to determine. Since it is no longer limited strictly to the judgements of disciplinary peers, the fear is that control will be weaker and result in lower quality work. Although the quality control process in Mode 2 is more broadly based, it does not follow that because a wider range of expertise is brought to bear on a problem that it will necessarily be of lower quality. It is of a more composite, multidimensional kind. [...]

from: Gibbons, Michael, Limoges, Camille, Nowotny, Helga, Schwartzman, Simon, Scott, Peter & Trow, Martin (1994), *The new production of knowledge*, Sage Publications

MICHEL CALLON

3 SOME ELEMENTS OF A SOCIOLOGY OF TRANSLATION

Domestication of the Scallops and the Fishermen of St. Brieuc Bay

I. Scallops and Fishermen

Highly appreciated by French consumers, scallops have only been systematically exploited for the last twenty years. In a short period, they have become a highly sought-after gourmandise to the extent that during the Christmas season, although prices are spectacularly high, sales increase considerably. They are fished in France at three locations: along the coast of Normandy, in the roadstead of Brest, and in St. Brieuc Bay. There are several different species of scallops. Certain ones, as in Brest, are coralled all year round. However, at St. Brieuc the scallops lose their coral during spring and summer. These characteristics are commercially important because, according to the convictions of the fishermen, the consumers prefer coralled scallops to those which are not.

Throughout the 1970s, the stock at Brest progressively dwindled due to the combined effects of marine predators (starfish), a series of hard winters which lowered the general temperature of the water, and the fishermen who, wanting to satisfy the insatiable consumers, dredged the ocean floor for scallops all year round without allowing time to reproduce. The production of St. Brieuc had also been falling off steadily during the same period, but fortunately the Bay was able to avoid the disaster. There were fewer predators and the consumers' preference for coralled scallops obliged the fishermen to stay on land for half the year. As a result of these factors, the reproduction of the stock decreased less in St. Brieuc Bay than at Brest.[1]

The object of this study is to examine the progressive development of new social relationships through the constitution of a "scientific knowledge" that occurred during the 1970s.[2] The story starts at a conference held at Brest in 1972. Scientists and the representatives of the fishing community assembled to examine the possibility of increasing the production of scallops by controlling their cultivation. The discussions were grouped around the following three elements.

1 Three researchers who are members of the Centre National d'Exploitation des Oceans (CNEXO)[3] have discovered during a

voyage to Japan that scallops are being intensively cultivated there. The technique is the following: the larvae are anchored to collectors immersed in the sea where they are sheltered from predators as they grow. When the shellfish attain a large enough size, they are "sown" along the ocean bed where they can safely develop for two or three years before being harvested. According to the researchers' accounts of their trip, this technique made it possible to increase the level of existing stocks. All the different contributions of the conference were focused around this report.

2 There is a total lack of information concerning the mechanisms behind the development of scallops. The scientific community has never been very interested in this subject. In addition, because the intensive exploitation of scallops had begun only recently, the fishermen knew nothing about the earlier stages of scallop development. The fishermen had only seen adult scallops in their dredges. At the beginning of the 1970s, no direct relationship existed between larvae and fishermen. As we will see, the link was progressively established through the action of the researchers.

3 Fishing had been carried out at such intensive levels that the consequences of this exploitation were beginning to be visible in St. Brieuc Bay. Brest had practically been crossed off the map. The production at St. Brieuc had been steadily decreasing. The scallop industry of St. Brieuc had been particularly lucrative and the fishermen's representatives were beginning to worry about the dwindling stock. The decline of the scallop population seemed inevitable and many feared that the catastrophe at Brest would also occur at St. Brieuc.

This was the chosen starting point for this paper. Ten years later, a "scientific" knowledge was produced and certified; a social group was formed (the fishermen of St. Brieuc Bay) through the privileges that this group was able to institute and preserve; and a community of specialists was organized in order to study the scallops and promote their cultivation. Basing my analysis on what I propose to call a *sociology of translation,* I will now retrace some part of this evolution and see the simultaneous production of knowledge and construction of a network of relationships in which social and natural entities mutually control who they are and what they want.

II. The Four Moments of Translation

To examine this development, we have chosen to follow an actor through his construction-deconstruction of nature and society. Our starting point here consists of the three researchers who returned from their voyage to the Far East. Where they came from and why they act is of little importance at this point of the investigation. They are the *primum movens* of the story analyzed here. We will accompany them during their first attempt at domestication. This endeavour consists of four moments which can in reality overlap. These moments constitute the different phases of a general process called translation, during which the identity of actors, the possibility of interaction, and the margins of manoeuvre are negotiated and delimited.

The Problematization, or How to Become Indispensable

Once they returned home, the researchers wrote a series of reports and articles in which they disclosed the impressions of their trip and the future projects they wished to launch. With their own eyes they had seen the larvae anchor themselves to collectors and grow undisturbed while sheltered from predators. Their question was simple: Is this experience transposable to France and, more particularly, to the Bay of St. Brieuc? No clear answer can be given because the researchers know that the *briochine (Pecten maximus)* is different from the species raised in Japanese waters (*Pecten patinopecten yessoeusis*). Since no one contradicts the researchers' affirmations, we consider their statements are held to be uncontestable. Thus the aquaculture of scallops at St. Brieuc raises a problem. No answer can be given to the following crucial question: Does *Pecten maximus* anchor itself during the first moments of its existence? Other questions which are just as important accompany the first. When does the metamorphosis of the larvae occur? At what rate do the young grow? Can enough larvae be anchored to the collectors in order to justify the project of restocking the bay?

But in their different written documents the three researchers did not limit themselves to the simple formulation of the above questions. They determined a set of actors[4] and defined their identities in such a way as to establish themselves an obligatory passage point in the network of relationships they were building. This double movement, which renders them indispensable in the network, is what I call *problematization.*

The Interdefinition of the Actors

The questions formed by the three researchers and the commentaries that they provided bring three other actors directly into the story: the scallops *(Pecten maximus)*; the fishermen of St. Brieuc Bay; and the scientific colleagues. The definitions of these actors, as they are presented in the scientists' report, are quite rough. However it is sufficiently precise to explain how these actors are necessarily concerned by the different questions which are formulated. These definitions as given by the three researchers themselves can be synthesized in the following manner.

1. *The fishermen of St. Brieuc:* they fish scallops to the last shellfish without worrying about the stock; they make large profits; if they do not slow down their zealous efforts, they will ruin themselves. However, these fishermen are considered to be aware of their long-term economic interests and, consequently, seem to be interested in the project of restocking the bay and approve of the studies which have been launched to achieve this plan. No other hypothesis is made about their identity. The three researchers make no comment about a united social group. They define an average fisherman as a base unit of a community which consists of interchangeable elements.

2. *Scientific colleagues:* participating in conferences or cited in different publications, they know nothing about scallops in general nor about those of St. Brieuc in particular. In addition, they are unable to answer the question about the way in which these shellfish anchor themselves. They are considered to be interested in advancing the knowledge which has been proposed. This strategy consists of studying the scallops in situ rather than in experimental tanks.

3. *The scallops of St. Brieuc:* a particular species *(Pecten maximus*) which everyone agrees is coralled only six months of the year. They have only been seen as adults, at the moment they are dredged from the sea. The question which is asked by the three researchers supposes that they can anchor themselves and will "accept" a shelter that will enable them to proliferate and survive.[5]

Of course, and without this the problematization would lack any support, the three researchers also reveal what they themselves are and what they want. They present themselves as "basic" researchers who, impressed by the foreign achievement, seek to advance the available knowledge concerning a species which had not been thoroughly studied before. By undertaking this investigation, these researchers hope to render the fishermen's life easier and increase the stock of scallops of St. Brieuc Bay.

This example shows that the problematization, rather than being a reduction of the investigation to a simple formulation, touches on elements, at least partially and locally, which are parts of both the social and the natural worlds. A single question—Does *Pecten maximus* anchor?—is enough to involve a whole series of actors by establishing their identities and the links between them.[6]

The Definition of Obligatory Passage Points (OPP)

The three researchers do not limit themselves simply to identifying a few actors. They also show that the interests of these actors lie in admitting the proposed research program. The argument which they develop in their paper is constantly repeated: if the scallops want to survive (no matter what mechanisms explain this impulse), if their scientific colleagues hope to advance knowledge on this subject (whatever their motivations may be), if the fishermen hope to preserve their long-term economic interests (whatever their reasons), then they must: (1) know the answer to the question, How do scallops anchor?, and (2) recognize that their alliance around this question can benefit each of them.

Figure 1 shows that the problematization possesses certain dynamic properties: it indicates the movements and detours that must be accepted as well as the alliances that must be forged. The scallops, the fishermen, and the scientific colleagues are fettered: they cannot attain what they want by themselves. Their road is blocked by a series of obstacles-problems. The future of *Pecten maximus* is threatened perpetually by all sorts of predators always ready to exterminate them; the fishermen, greedy for short-term profits, risk their long-term survival; scientific colleagues who want to develop knowledge are obliged to admit the lack of preliminary and indispensable observations of scallops in situ. As for the three researchers, their entire project turns around the question of the anchorage of *Pecten maximus.* For these actors the alternative is

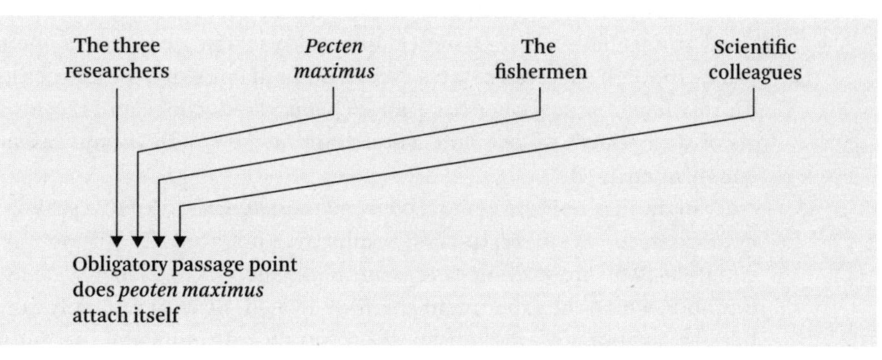

Figure 1

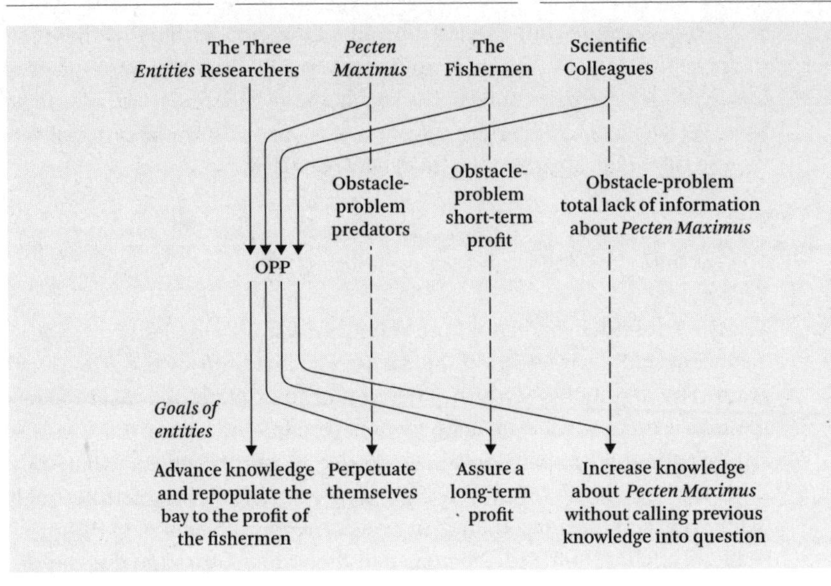

Figure 2

clear; either one changes direction or one recognizes the need to study and obtain results about the way in which larvae anchor themselves.[7]

As Figure 2 shows, the problematization describes a system of alliances, or associations, between entities, thereby defining the identity and what they "want." In this case, a holy alliance must be formed in order to induce the scallops of St. Brieuc Bay to multiply.

The Devices of Interessement, or How the Allies are Locked into Place

We have emphasized the hypothetical aspect of the problematization. On paper, or more exactly, in the reports and articles presented by the three researchers, the identified groups have a real existence. But reality is a process. Like a chemical body, it passes through successive states. At this point in our story, the entities identified and the relationships envisaged have not yet been tested. The scene is set for a series of trials of strength whose outcome will determine the solidity of our researchers' problematization. [...]

In the beginning, a general consensus existed: the idea that scallops anchor was not discussed.[8] However, the first results were not accepted without preliminary negotiations. The proposition: "*Pecten maximus* anchors itself in its larval state" is an affirmation which the experiments performed at St. Brieuc eventually called into question. No anchorages were observed on certain collectors and the number of larvae which anchored on the collectors never attained the Japanese levels. At what number can it be confirmed and accepted that scallops, in general, do anchor themselves? The three researchers are prepared for this objection because in their first communication they confirm that the observed anchorages did not occur accidentally: it is here that we see the importance of the negotiations which were carried out with the scallops in order to increase the *interessement* and of the acts of enticement which were used to retain the larvae (horsehair rather than nylon, and so on). With scientific colleagues, the transactions were simple: the discussion of the results shows that they were prepared to believe in the principle of anchorage and that they judged the experiment to be convincing. The only condition that the colleagues posed is that the existence of previous work be recognized, work that had predicted, albeit imperfectly, the scallops' capacity to anchor.[9] It is at this price that the number of anchorages claimed by the researchers will be judged as sufficient. Our three researchers accept, after ironically noting that all bonafide discoveries miraculously unveil precursors, who had been previously ignored.[10]

Transactions with the fishermen, or rather, with their representatives, are nonexistent. They watch like amused spectators and wait for the final verdict. They are prepared simply to accept the conclusions drawn by the specialists. Their consent is obtained (in advance) without any discussion.

Therefore for the most part, the negotiation is carried between three parties since the fourth partner was enrolled without any resistance. This example illustrates the different possible ways in which the actors are enrolled: physical violence (against the predators), seduction, transaction, and consent without discussion. This example mainly shows that the definition and distribution of roles (the scallops which anchor themselves, the fishermen who are persuaded that the collectors could help restock the bay, the colleagues who believe in the anchorage) are a result of multilateral negotiations during which the identity of the actors is determined and tested.

The Mobilization of Allies: Are the Spokesmen Representative?

Who speaks in the name of whom? Who represents whom? These crucial questions must be answered if the project led by the researchers is to succeed. This is because, as with the description of *interessement* and enrollment, only a few rare individuals are involved, whether these be scallops, fishermen or scientific colleagues. [. . .]

First, the notion of translation emphasizes the continuity of the displacements and transformations which occur in this story: displacements of goals and interests and also displacements of devices, human beings, larvae, and inscriptions. Because of a series of unpredictable displacements, all the processes can be described as a translation which leads all the actors concerned to pass, through various metamorphoses and transformations, by the three researchers and their development project.

To translate is to displace: the three untiring researchers attempt to displace their allies to make them pass by Brest and their laboratories. But to translate is also to express in one's own language what others say and want, why they act in the way they do and how they associate with each other: it is to establish oneself as a spokesman. At the end of the process, if it is successful, only voices speaking in unison will be heard. The three researchers talk in the name of the scallops, the fishermen, and the scientific community. At the beginning these three universes were separate and had no means of communication with one another. At the end a discourse of certainty has unified them, or, rather, has brought them into a relationship with one another in an intelligible manner. But this would not have been possible without the different sorts of displacements and transformations presented above, the negotiations, and the adjustments that accompanied them. To designate these two inseparable mechanisms and their result, we use the word *translation.* The three researchers translated the fishermen, the scallops, and the scientific community.

Translation is a process before it is a result. That is why we have spoken of moments which in reality are never as distinct as they are in this paper. Each of them marks a progression in the negotiations which results in the designation of the legitimate spokesmen who, in this case study, say what the scallops want and need and are not disavowed: the problematization, which was only a simple conjecture, was transformed into mobilization. Dissidence plays a different role since it brings into question some of the gains of the previous stages. The displacements and the spokesmen are challenged or refused. The actors implicated do not acknowledge their roles in this story nor the slow drift in which they had participated, in their opinion, wholeheartedly. As the aphorism says, "traduttore-traditore," from translation to treason there is only a short step. It is this step that is taken in the last stage. New displacements take the place of the previous ones but these divert the actors from the obligatory passage points that had been imposed upon them. New spokesmen are heard that deny the representativity of the previous ones. Translation continues but the equilibrium has been modified.

Translation is the mechanism by which the social and natural worlds progressively take form. The result is a situation in which certain entities control others. Understanding what sociologists generally call power relationships means describing the way in which actors are defined, associated, and simultaneously obliged to remain faithful to their alliances. The repertoire of translation is not only designed to give a symmetrical and tolerant description of a complex process which constantly mixes together a variety of social and natural entities. It also permits an explanation of how a few obtain the right to express and to represent the many silent actors of the social and natural worlds they have mobilized.

from: Biagioli, Mario, ed. (1999): The Science Studies Reader, New York.

NOTES

1. The notion of "stock" is widely used in population demography. In the present case the stock designates the population of scallops living and reproducing in St. Brieuc Bay. A given stock is designated by a series of parameters that vary over time: overall number, cohorts, size, natural mortality rate, rate of reproduction, and so on. Knowledge of the stock thus requires systematic measures which make it possible to forecast changes. In population dynamics mathematical models define the influence of a range of variables (e.g., intensity of fishing and the division of catch between cohorts) upon the development of the stock. Population dynamics is thus one of the essential tools for what specialists in the study of maritime fishing call the rational management of stocks.
2. For this study we had available all the articles, reports, and accounts of meetings that related to the experiments at St, Brieuc and the domestication of scallops. About twenty interviews with leading protagonists were also undertaken.
3. Centre National d'Exploitation des Océans (CNEXO) is a public body that was created in the early 1970s to undertake research designed to increase knowledge and means of exploiting marine resources.
4. The term *actor* is used in the way that semioticians use the notion of the *actant* (Greimas and Courtes 1979). For the implication of external actors in the construction of scientific knowledge or artifacts see the way in which Pinch and Bijker (1984) make use of the notion of a social group. The approach proposed here differs from this in various ways: first, as will be suggested below, the list of actors is not restricted to social entities; but second, and most important, because the definition of groups, their identities and their wishes are all constandy negotiated during the process of translation. Therefore, these are not pregiven data but take the form of an hypothesis (a problematization) that is introduced by certain actors and is subsequently weakened, confirmed, or transformed.
5. The reader should not impute anthropomorphism to these phrases! The reasons for the conduct of scallops—whether these lie in their genes, in divinely ordained schemes, or anything else—matter little! The only thing that counts is the definition of their conduct by the various actors identified. The scallops are deemed to attach themselves just as fishermen are deemed to follow their short-term economic interests. They therefore act.
6. On the negotiable character of interests and identities of the actors see Callon (1980).
7. As can be discerned from its etymology, the word *problem* designates obstacles that are thrown across the path of an actor and which hinder his movement. This term is thus used in a manner which differs entirely from that current in the philosophy of science and epistemologa Problems are not spontaneously generated by the state of knowledge or by the dynamics of progress in research. Rather they result from the definition and interrelation of actors that were not previously linked to one another. To problematize is simultaneously to define a series of actors and the obstacles which prevent them from attaining the goals or objectives that have been imputed to them. Problems, and the postulated equivalences between them, result from the interaction between a given actor and all the social and natural entities which it defines and for which it seems to become indispensable.
8. The discussions were recorded in reports which were made available.
9. One participant in the discussion, commenting on the report of Buestel et al., noted: "At a theoretical level we must not minimise what we know already about scallops. . . . It is important to remember that the biology of *Pecten* was somewhat better known than you suggested."
10. Buestel et al. Resultats préliminaires.

REFERENCES

Callon, Michel. (1980) "Struggles and negotiations to define what is problematic and what is not: the socio-logic of translation." in *The Social Process of Scientific Investigation. Sociology of the Sciences Yearbook, Vol. 4,* ed. K. D. Knorr and A. Cicourel, Boston: D. Reidel Publishing Company.

Callon, Michel. (1986) "The sociology of an actor-network" in *Mapping the Dynamics of Science and Technology,* ed. M. Callon, J. Law, and A. Rip, London: Macmillan.

Callon, Michel, J. Law, and A. Rip, eds. (1986) *Mapping the Dynamics of Science and Technology*, London: Macmillan.

Greimas, A J., and J. Courtes. (1979) *Sémiotique: dictionnaire raisonné de la théorie du langage*, Paris: Hachette.

Latour, Bruno. 1987, *Science in Action.* Milton Keynes: Open University Press.

Pinch, T, J., and W, Bijker. (1984) "The social construction of facts and artefacts: or how the sociology of science and the sociology of technology might benefit each other.", *Social Studies of Science* 14: 388-441.

Karin Knorr Cetina

4 WHAT IS A LABORATORY?

[. . .]

Much of the literature on the history and methodology of science relies on the notion of the experiment as the basic unit of analysis. I want to suggest in this chapter how the notion of the laboratory—beyond its identification as just the physical space in which experiments are conducted—has emerged historically as a set of differentiated social and technical forms, carrying systematic weight in our understanding of science. The importance of this concept is linked to the reconfiguration of both the natural and social orders that, I will argue, constitutes laboratories in crucial ways. Further, I will argue that these reconfigurations work quite differently in different fields of science, generating different cultural, social, and technical stances.

2.1 Laboratories as Reconfigurations of Natural and Social Orders

I want to begin by proposing that laboratories provide an "enhanced" environment that "improves upon" natural orders in relation to social orders. How does this improvement come about? The studies we have of laboratory work (e.g., Latour and Woolgar 1979; Knorr 1977; Knorr Cetina 1981; Zenzen and Restivo 1982; Lynch 1985; Giere 1988; Gooding et al. 1989; Pickering 1995) imply that it rests upon the *malleability* of natural objects. Laboratories are based upon the premise that objects are not fixed entities that have to be taken "as they are" or left by themselves. In fact, one rarely works in laboratories with objects as they occur in nature. Rather, one works with object images or with their visual, auditory, or electrical traces, and with their components, their extractions, and their "purified" versions. There are at least three features of natural objects a laboratory science does not have to accommodate: first, it does not need to put up with an object *as it is,* it can substitute transformed and partial versions. Second, it does not need to accommodate the natural object *where it is,* anchored in a natural environment; laboratory sciences bring objects "home" and manipulate them on their own terms, in the laboratory. Third, a laboratory science need not accommodate an event *when it happens;* it can dispense with natural cycles of occurrence and make events happen frequently enough for continuous study. Of course, the history of science is also a history of lost opportunities and varying successes in accomplishing these transitions. But it should be clear that not having to confront objects within their natural orders is epistemically advantageous for the pursuit of science; laboratory practice entails the detachment of objects from their natural environment and their installation in a new phenomenal field defined by social agents. [...]

My point is that scientists have been similarly shaped and transformed with regard to the kind of agents and processing devices they use in inquiry. Just as objects are transformed into images, extractions, and a multitude of other things in laboratories, so are scientists reconfigured to become specific epistemic subjects. As we shall see later, the scientist who acts as a bodily measurement device (by hearing and seeing signals) is also present in molecular biology (Chapter 4). By the time the reconfigurations of self-other-things that constitute laboratories have taken place, we are confronted with a newly emerging order that is neither social nor natural—an order whose components have mixed genealogies and continue to change shape as laboratory work continues.

2.2 From Laboratory to Experiment

What I have said so far refers to laboratory processes in general. I have neglected the fact that concrete laboratory reconfigurations are shaped in relation to the kind of work that goes on within the laboratory. This is where experiments come into the picture; through the technology they employ, experiments embody and respond to reconfigurations of natural and social orders.

Let me draw attention to three different types of laboratories and experiments in the contemporary sciences of particle physics, molecular biology, and the social sciences. In distinguishing between these types, I shall take as my starting point the constructions placed upon natural objects in these areas of science and their embodiment in the respective technologies of experimentation. I want to show how, in connection with these different constructions, laboratories and experiments become very different entities and enter very different kinds of relationships with each other. First, laboratories and experiments can encompass more or less distinctive and independent activities; they can be assembled into separate types, which confront and play upon each other, or disassembled to the extent that they appear to be mere aspects of one another. Second, the relationship between local scientific practices and "environments" also changes as laboratories and experiments are differently assembled. In other words, reconfigurations of natural and social orders can in fact *not* be entirely contained in the laboratory space. Scientific fields are composed of more than one laboratory and more than one experiment; the reconfigurations established in local units have implications for the kind of relationship that emerges between these units, and beyond.

In the following, I shall do no more than outline some of these issues in a most cursory manner. I shall thereby introduce a first *set of differences* between the sciences of molecular biology and high energy physics, which will be the focus of much detail in later chapters. In this section, I want to draw attention to the diverse *meanings* of "experiment" and "laboratory" that are indicated in different reconfigurations (see also

Hacking 1992b). I want to indicate the varying significance of laboratories and experiments in relation to each other in three situations, which I distinguish in terms of whether they use a technology of correspondence, a technology of treatments and interventions, or a technology of representation (see also Shapin and Schaffer 1985). The construction placed upon the objects of research varies accordingly; in the first case, objects in the laboratory *stage* real-world phenomena; in the second, they are *processed partial versions* of these phenomena; in the third, they are *signatures* of the events of interest to science. Note that the distinctions drawn are not meant to point to some "essential" differences between fields; rather, they are an attempt to capture how objects are primarily featured and attended to in different areas of research. To illustrate the differences and to emphasize the continuity between mechanisms at work within science and outside, I shall first draw upon examples of "laboratories" and "experimentation" from outside natural science—the psychoanalyst's couch, the medieval cathedral, and the war game.

2.2.1 Experiments (Almost) Without Laboratories: Objects That Stage Real-Time Events

I begin with the war game. The hallmark of a *Kriegsspiel* in the past was that it took place on a "sand table," a kind of sandbox on legs, in which the geographic features of a potential battle area were built and whole battles were fought by toy armies. The landscape was modeled after the scene of a real engagement in all relevant respects, and the movements made by the toy soldiers corresponded as closely as possible to the expected moves of real armies. The sandbox war game was an eighteenth-century invention and was developed further by Prussian generals. Its modern equivalent is the computer simulation, which has become widely used not only in the military but in many areas of science, where real tests are impracticable for one reason or another. Computer simulations are also increasingly used in laboratory sciences to perform experiments. Indeed, the computer has been called a laboratory (e.g., Hut and Sussman 1987); it provides its own "test-bench" environment.

The point here is that many real-time laboratory experiments bear the same kind of relationship to reality as the war game bears to real war or the computer simulation bears to the system being modeled; they *stage* the action. As an example, consider most experiments in the social sciences, particularly in social psychology, or in economics, in research on problem solving and the like. To illustrate, in experimental research on decision making by juries, research participants (often college students) are set up in the way real juries would be in court. They are given information on a case and asked to reach a verdict in ways that approximate real jury decision processes. They may even be exposed to pleas by the mock accused and other elements of real-time situations (e.g., MacCoun 1989). Research on the heuristics of problem solving uses a similar design. Experts, lay persons, or novices to an area are recruited and asked to search for a solution to a simulated problem (Kahneman, Slovic, and Tversky 1982). One difference from the war game in the sandbox is that the experimental subjects in the social sciences are not toys but members of the targeted population. For example, they may be real experts who play experts in the laboratory, or students who are thought to be representative of the jury pool. Nonetheless, social science experiments receive the same kind of criticism as computer simulations do. While the subjects recruited for the experiment may not differ much from the persons about whom results are to be generated, the setting is artificial, and the difference this makes, with respect to the behavior generated in experimental situations, is poorly understood. What the critics question is whether generalizable results can be reached by studying behavior in mock settings when the factors distinguishing the simulation from real-time events are not known or have not been assessed.

Researchers in these areas are, of course, aware of this potential source of error. As a consequence, they take great care to design experimental reality so that in all relevant respects they come close to the perceived real-time processes. In other words, they develop and deploy a *technology of correspondence.* For example, they set up a system of assurance through which correct correspondence with the world is monitored. One outstanding characteristic of this system of assurance is that it is based on a theory of nonintervention. In "blind" and "doubleblind" designs, researchers attempt to eradicate the very possibility that they will influence experimental outcomes. In fact, experimental designs consist of implementing a world simulation, on the one hand, and implementing a thorough separation between the experimental subjects and the action, interests, and interpretations of the researchers, on the other.

Now consider the laboratory in social science areas. It does not, as a rule, involve a richly elaborated space—a place densely stacked with instruments and materials and populated by researchers. In many social sciences, the laboratory is reduced to a room with a one-way mirror that includes perhaps a table and some chairs. In fact, experiments may be conducted in researchers' offices when a one-way mirror is not essential. But even when a separate laboratory space exists, it tends to be used only when an experiment is conducted, which, given the short duration of such experiments, happens only rarely. The laboratory is a virtual space and, in most respects, co-extensive with the experiment. Like a stage on which plays are performed from time to time, the laboratory is a *storage room* for the *stage props* that are needed when social life is instantiated through experiments. The "objects" featured on the stage are *players of the social form.* The hallmark of their reconfiguration seems to be that they are called upon to perform everyday life in a competent manner, and to behave under laboratory conditions true to the practice of real-time members of daily life. [...]

If we now turn to the laboratories where the manipulation takes place, it should come as no surprise that they are not, as in the first case, storage rooms for stage props. It seems that it is precisely with the above-mentioned processing approach, the configuration of objects as materials to be interfered with, that laboratories "come of age" and are established as distinctive and separate entities. What kind of entities? Take the classical case of a bench laboratory as exemplified in molecular biology (see also Lynch 1985; Jordan and Lynch 1992; Fujimura 1987; Amann 1990, 1994). This bench laboratory is always activated; it is an actual space in which research tasks are performed continuously and simultaneously. The laboratory has become a *workshop* and a *nursery* with specific goals and activities. In the laboratory, different plant and animal materials are maintained, bred, nourished, warmed, observed, and prepared for experimental manipulation. Surrounded by equipment and apparatus, they are used as technical devices in producing experimental effects.

The laboratory is a repository of processing materials and devices that continuously feed into experimentation. More generally, laboratories are objects of work and attention over and above experiments. Laboratories employ caretaking personnel for the sole purpose of tending to the waste, the used glassware, the test animals, the apparatus, and the preparatory and maintenance tasks of the lab. Scientists are not only researchers but also caretakers of the laboratory. As will become clearer in later chapters, certain types of tasks become of special concern to heads of laboratories, who tend to spend much of their time representing and promoting "their" lab (see Chapter 9). In fact, laboratories are also social and political structures that "belong" to their heads in the sense that they are attributed to them and identified with them. Thus, the proliferation of laboratories as objects of work is associated with the emergence of a two-tier system of social organization of agents and activities—the lab level and the experimental level. Experiments, however, tend to have little unity. In fact, they appear to be dissolved into processing activities, parts of which are occasionally pulled together for the purpose of publication. As laboratories gain symbolic distinctiveness and become a focus of activities, experiments lose some of the "wholeness" they display in social scientific fields. When the laboratory becomes a permanent facility, experiments can be continuous and parallel, and they even begin to blend into one another. Thus, experiments dissolve into experimental work, which, in turn, is continuous with laboratory-level work.

But there is a further aspect to the permanent installation of laboratories as *internal processing environments*. This has to do with the phenomenon that laboratories now are collective units that encapsulate within them a traffic of substances, materials, equipment, and observations. Phrased differently, the laboratory houses within it the circuits of observation and the traffic of experience that medieval cathedral builders brought about through travel. At the same time, neither the traffic of specimens and materials nor the system of surveillance is solely contained in the laboratory. If the laboratory has come of age as a continuous and bounded unit that encapsulates internal environments, it has also become a participant in a larger field of communication and mutual observation. the traffic of objects, researchers, and information produces a *lifeworld* within which laboratories are locales, but which extends much further than the boundaries of single laboratories.

[. . .]

2.3 Some Features of the Laboratory Reconsidered

I have argued that the notion of a laboratory in recent sociology of science is more than a new field of exploration, a site which houses experiments (Shapin 1988), or a locale in which methodologies are put into practice. I have associated laboratories with the notion of reconfiguration, with the setting-up of an order in laboratories that is built upon upgrading the ordinary and mundane components of social life. Laboratories *recast* objects of investigation by inserting them into new temporal and territorial regimes. They play upon these objects' natural rhythms and developmental possibilities, bring them together in new numbers, renegotiate their sizes, and redefine their internal makeup. They also invent and recreate these objects from scratch (think of the particle decays generated by particle colliders). In short, they create new configurations of objects that they match with an appropriately altered social order.

In pointing out these features I have defined laboratories as *relational* units that gain power by instituting *differences* with their environment: differences between the reconfigured orders created in the laboratory and the conventions and arrangements found in everyday life, but of course also differences between contemporary laboratory setups and those found at other times and places. Laboratories, to be sure, not only play upon the social and natural orders as they are experienced in everyday life. They also play upon themselves; upon their own previous makeup and at times upon those of competing laboratories. What I said in Sections 2.1 and 2.2 implies that one can link laboratories as *relational* units to at least three realities: to the environment they reconfigure, to the experimental work that goes on within them and is fashioned in terms of these reconfigurations, and to the "field" of other units in which laboratories and their features are situated.

Laboratories introduce and utilize *specific differences* between processes implemented in them and processes in a scientific field. Take the case of the space telescope mentioned earlier, or the recently developed underwater telescope. The underwater telescope does not operate in space or on mountaintops, but three miles beneath the ocean. Unlike previous telescopes, it does not observe electromagnetic radiation but streams of neutrino particles thought to be emitted by components of distant galaxies, such as black holes. Neutrinos are an elusive type of particle that travels easily through the earth and through space, undeterred by cosmic obstacles. The water under which the telescope is built serves as a screening system that filters out unwanted high energy particles that might mask the neutrino signals. Since even several miles of water do not offer enough shielding, however, the telescope, looking *down* rather than up, picks up signals from particles that fly all the way through the earth and emerge from the ocean floor.

The notion of reconfiguration needs to be extended to include issues continuously at stake in laboratories: the ongoing work of instituting specific differences from which epistemic dividends can be derived, and the work of boundary maintenance with regard to the natural and everyday order (see also Gieryn 1983). We need to conceive of laboratories as processes through which reconfigurations are negotiated, implemented, superseded, and replaced. Doing so would imply a notion of *stages* of laboratory processes, which can be historically investigated and which may also be important for questions of consensus formation (see Shapin 1988; Hessenbruch 1992; Lachmund 1997; Giere 1988). But it also implies that we have to expect different *types* of laboratory processes in different areas, resulting from cumulative processes of differentiation. [...]

from: Knorr Cetina, Karin (1999), *Epistemic Cultures: How the Sciences Make Knowledge*, Cambridge, MA: Harvard University Press,

Hans-Jörg Rheinberger

5 EXPERIMENTAL SYSTEMS AND EPISTEMIC THINGS

> [The] meticulous care required to connect things in unbroken succession.
>
> —Goethe, "The Experiment As Mediator between Object and Subject"

At a symposium on the structure of enzymes and proteins in 1955, Paul Zamecnik read a paper on the "Mechanism of Incorporation of Labeled Amino Acids into Protein." When, in the ensuing discussion, Sol Spiegelman reported his own experiments on the induction of enzymes in yeast cultures, Zamecnik responded, "we would like to study induced enzyme formation, too; but that reminds me of a story Dr. Hotchkiss told me of a man who wanted to use a new boomerang but found himself unable to throw his old one away successfully."[1]

What Does it Mean to Do Experiments?

Better than any lengthy description, the opening anecdote illustrates an essential feature of experimental practice. It expresses an experience familiar to every working scientist: the more he or she learns to handle his or her own experimental system, the more it plays out its own intrinsic capacities. In a certain sense, it becomes independent of the researcher's wishes just because he or she has shaped it with all possible skill. What Lacan states for the structuralist human sciences holds here, too: "The subject is, as it were, internally excluded from its object."[2] It is this "intimate exteriority," or "extimacy,"[3] captured in the image of a boomerang, that we may call virtuosity.

Virtuosity creates pleasure. When Alan Garen once asked Alfred Hershey for his idea of scientific happiness, he answered: "To have one experiment that works, and keep doing it all the time." As Seymour Benzer wrote later, this became known as "Hershey Heaven" among the first generation of molecular biologists.[4]

In his autobiography, François Jacob has formulated the same experience from the perspective of being engaged in an ongoing research process:

> In analyzing a problem, the biologist is constrained to focus on a fragment of reality, on a piece of the universe which he arbitrarily isolates to define certain of its parameters. In biology, any study thus begins with the choice of a "system." On this

> choice depend the experimenter's freedom to maneuver, the nature of the questions he is free to ask, and even, often, the type of answer he can obtain.[5]

Thus, we have, at the basis of biological research, the choice of a system, and a range of maneuvers that it allows us to perform. Which is, if I see it correctly, a specific reformulation of Heidegger's claim that to "open up a sphere," and to "establish a procedure" is what modern research, considered as representing the "essence" of occidental science, is all about.

In his "The Age of the World Picture," Heidegger states with respect to the modern sciences:

> The essence of what we today call science is research. [But] in what does the essence of research consist? In the fact that knowing establishes itself as a procedure within some realm of what is, in nature or in history. Procedure does not mean here merely method or methodology. For every procedure already requires an open sphere *(offener Bezirk)* in which it moves. And it is precisely the opening of such a sphere that is the fundamental event in research.[6]

To open a sphere and to establish a procedure: such ought to be the grounding feature of the modern sciences, viewed from a Heideggerian point of view.

With respect to intent and context, these quotations are utterly different. Zamecnik, Garen, Jacob, and Heidegger speak about experimentation in the light of acquaintance, satisfaction, constraint, and conquest. But in another respect they coincide: they all identify a research setting, or experimental system, as the core structure of scientific activity. Such a view, if taken seriously, entails epistemological as well as historiographical consequences. If we accept the thesis that *research* is the basic procedure of the modern sciences, we are invited to explore how research gets enacted at the frontiers between the known and the unknown. If we accept that biological research in particular begins with the choice of a system rather than with the choice of a theoretical framework, it will be in order to focus attention on the characterization of experimental systems, their structure, and their dynamics. To speak of the "choice" of a system here does not mean that such arrangements are there from the beginning. To arrive at an experimental system is itself a laborious process, as my case study of the group at MGH will show. My emphasis is on the materialities of research. Therefore, as my point of departure I will not directly address the theory and practice issue and the relation between theory and practice, the theory-ladenness of observation, or the underdetermination of theory by experiment. My approach tries to escape this "theory first" type of philosophy of science perspective. For want of a better term, the approach I am pursuing might be called "pragmatogonic." I would like to convey a sense of what it means for the participants in the endeavor to be engaged in epistemic practices, that is, in irrevocably experimental situations. Here I claim, with Frederick Holmes, "it is the investigations themselves which are at the heart of the life of an active experimental scientist. For him ideas go in and come out of investigations, but by themselves are mere literary exercises. [I]f we are to understand scientific activity at its core, we must immerse ourselves as fully as possible into those investigative operations."[7]

In this chapter, I first turn to some structural characteristics of such investigative operations on the level of relatively *longue durée.* Let me recall an episode from the end of the eighteenth century. When Goethe was performing the optical experiments that led to his theory of colors, he wrote, in 1793, a remarkable essay entitled "The Experiment as Mediator between Object and Subject."[8] In this essay, Goethe addresses his problem in a similar, but still different, vein, neither with respect to virtuosity nor to pleasure, but—conforming to what Friedrich Kitder has called the "Aufschreibesystem 1800"[9]—with respect to the duty of the scientist. The central sentence reads as follows: "To follow every single experiment through its variations is the real task of the scientific researcher." Goethe compares what he calls "Versuch" with a point from which light is emitted in all possible directions. Through the step-by-step exploration of all of them, a research network is built up that eventually will come into contact with neighboring networks. Establishing such fields, according to Goethe, is the primary task of the experimentalist; disciplinary junctures may be the final outcome of his endeavor. "Thus when we have done an experiment of this type, found this or that piece of empirical evidence, we can never be careful enough in studying what lies next to it or derives directly from it. This investigation should concern us more than the discovery of what is related to it."[10] Five years later, Goethe asked Schiller to comment.[11] In his reply, Schiller immediately pointed to the core of the argument: "It is quite obvious to me how dangerous it is to try to demonstrate a theoretical proposition directly by experiments."[12]

Experimental Systems

According to a long-standing tradition in philosophy of science, experiments have been seen as singular, well-defined empirical instances embedded in the elaboration of a theory and performed in order to corroborate or to refute certain hypotheses. In the classical formulation of Karl Popper, "the theoretician puts certain definite questions to the experimenter, and the latter, by his experiments, tries to elicit a decisive answer to these questions, and to no others. All other questions he tries hard to exclude."[13] Despite the radical shift in perspective in which social studies of science have attempted to deny the naked experiment its ability to decide scientific controversies, the familiar notion of the experiment as a test of a hypothesis is still virulent in them. Even Harry Collins's argument from the "experimenter's regress" embraces, in its very rejection, a view of the experiment as an ultimate arbiter.[14]

What does it mean to speak of experimental systems, in contrast to this clear-cut rationalist picture of experimentation as a theory-driven activity? Ludwik Fleck, Popper's long neglected contemporary, has drawn our attention to the manufacture of scientific practices in twentieth-century biomedical sciences and has argued that—contrary to Popper's claim—scientists usually do not deal with single experiments in the context of a properly delineated theory. "Every experimental scientist knows just how little a single experiment can prove or convince. To establish proof, an entire *system of experiments* and controls is needed, set up according to an assumption or style and performed by an expert."[15] A researcher thus does not, as a rule, deal with isolated experiments in relation to a theory, but rather with a whole experimental arrangement designed to produce knowledge that is not yet at his disposal. What is even more important, the experimental scientist deals with systems of experiments that usually are not well defined and do not provide clear answers. Fleck even goes so far as to claim that "if a research experiment were well defined, it would be altogether unnecessary to perform it. For the experimental arrangements to be well defined, the outcome must be known in advance; otherwise the procedure cannot be limited and purposeful."[16] These remarks are not to be taken as a trivial characterization of a de facto imperfection of a particular research activity. They are to be taken as a profound reorientation of our view of the inner workings of this process, a process "driven from behind,"[17] a genuinely polysemic procedure defined by ambiguity, not one just limited by finite precision.

Experimental systems are to be seen as the smallest integral working units of research. As such, they are systems of manipulation designed to give unknown answers to questions that the experimenters themselves are not yet able clearly to ask. Such setups are, as Jacob once put it, "machines for making the future."[18] They are not simply experimental devices that generate answers; experimental systems are vehicles for materializing questions. They inextricably cogenerate the phenomena or material entities and the concepts they come to embody. Practices and concepts thus "come packaged together."[19] A single experiment as a crucial test of a properly delineated conception is not the simple, elementary, or basic situation of the experimental sciences. The inverse holds. Any simple case is the "degeneration" of an elementarily complex experimental situation. As Bachelard reminds us, "simple always means simplified. We cannot use simple concepts correctly until we understand the process of simplification from which they are derived."[20] It is only in the process of making one's way through a complex experimental landscape that scientifically meaningful simple things get delineated; in a non-Cartesian epistemology, they are not given from the beginning. They are the inescapably historical product of a purification procedure.[21] This is, again and again, the experience we find when we look at autobiographical science narratives.[22] But this is also what we find when we try to follow particular cases in the history of the modern life sciences. One of them will be expounded in Chapter 3 and traced throughout the rest of the book.

Epistemic Things, Technical Objects

In inspecting experimental systems more closely, two different yet inseparable elements can be discerned.[23] The first I call the research object, the scientific object, or the "epistemic thing." They are material entities or processes—physical structures, chemical reactions, biological functions—that constitute the objects of inquiry. As epistemic objects, they present themselves in a characteristic, irreducible vagueness. This vagueness is inevitable because, paradoxically, epistemic things embody what one does not yet know. Scientific objects have the precarious status of being absent in their experimental presence; they are not simply hidden things to be brought to light through sophisticated manipulations. A mixture of hard and soft, like Serres's veils, they are "object, still, sign, already; sign, still, object, already."[24] With Bruno Latour, we can claim it to be characteristic for the sciences in action that "the new object, at the time of its inception, is still undefined. [At] the time of its emergence, you cannot do better than explain what the new object is by repeating the list of its constitutive actions. [The] proof is that if you add an item to the list you *redefine the object,* that is, you give it a new shape."[25]

To enter such a process of operational redefinition, one needs an arrangement that I refer to as the experimental conditions, or "technical objects." It is through them that the objects of investigation become entrenched and articulate themselves in a wider field of epistemic practices and material cultures, including instruments, inscription devices, model organisms, and the floating theorems or boundary concepts attached to them. It is through these technical conditions that the institutional context passes down to the bench work in terms of local measuring facilities, supply of materials, laboratory animals, research traditions, and accumulated skills carried on by long-term technical personnel. In contrast to epistemic objects, these experimental conditions tend to be characteristically determined within the given standards of purity and precision. The experimental conditions "contain" the scientific objects in the double sense of this expression: they embed them, and through that very embracement, they restrict and constrain them.[26] Superficially, this constellation looks simple and obvious. But the point to be made is that within a particular experimental system both types of elements are engaged in a nontrivial interplay, intercalation, and interconversion, both in time and in space. The technical conditions determine the realm of possible representations of an epistemic thing; and sufficiently stabilized epistemic things turn into the technical repertoire of the experimental arrangement.

Take the following example, to which I will return in detail in Chapter 13: When Heinrich Matthaei and Marshall Nirenberg, in their bacterial in vitro system of protein synthesis, introduced synthetic polyuridylic acid, among other ribonucleic acids, as a possible template for polypeptide formation, the genetic code assumed the quality of an experimental epistemic thing. When the genetic code was solved, the polyuridylic acid assay, within the same in vitro system, was turned into a subroutine for the functional elucidation of the protein synthesizing organelles, the ribosomes. To add one more example, less than twenty years ago, enzymatic sequencing of DNA was a scientific object par excellence. It was a new possible mode of primary structure determination among older ones.[27] A few years later, it became a procedure that had been adopted by the leading DNA laboratories around the world. In the early 1980s, it was transformed into a technical object with all the characteristics of such a "translation." Today, every biochemical laboratory may order a sequence kit, including buffers, nucleotides, and enzymes from a biochemical company, and perform the sequence reaction routinely in a semi-automatic machine. Latour has spoken of "black boxing" in this context.[28] Unfortunately, this expression mainly reflects one particular aspect of the process: its "routine" nature after the event. Perhaps at least as important, however, is its impact on a new generation of emerging epistemic things. Black boxing does not mean just setting aside.

Through this kind of recurrent determination, certain sets of experiments become clearer in some directions but at the same time less independent because they more and more rely on a hierarchy of established procedures. "Once a field has been sufficiently worked over so that the possible conclusions are more or less limited to existence or nonexistence, and perhaps to quantitative determination, the experiments will become increasingly better defined. But they will no longer be independent, because they are carried along by a system of earlier experiments and decisions."[29]

The difference between experimental conditions and epistemic things, therefore, is functional rather than structural. We cannot once and for all draw such a distinction between different components of a system. Whether an object functions as an epistemic or a technical entity depends on the place or "node" it occupies in the experimental context. Despite all possible degrees of gradation between the two extremes, which leave room for all possible degrees of hybrids between them, their distinctness is clearly perceived in scientific practice. It organizes the laboratory space with its messy benches and specialized local precision services as well as the standard scientific text with its specialized sections on "materials and methods" (technical things), "results" (halfway-hybrids) and "discussion" (epistemic things).[30]

If both types of entities are engaged in a relation of exchange, of blending and mutual transformation, why then not cancel the distinction altogether? Does it not simply perpetuate the traditional, problematic distinction between basic research and applied science, between science and technology? If science in action should not be conceived in terms of an asymmetric relation from theory to practice, why then uphold a gradient between epistemic and technical objects? Why then construct a division whose only effect is that it permanently has to be undone? The answer is: because it helps to assess the game of innovation, to understand the occurrence of unprecedented events and with that, the essence of research. [...]

from: Rheinberger, Hans-Jörg (1997) *Toward a History of Epistemic Things. Synthesizing Proteins in the Test Tube*,

NOTES

1 Zamecoit, Keller, Littlefield, Hoagland, and Loftfield 1956. The symposium was held at the Research Conference for Biology and Medicine of the Atomic Energy Commission, Oak Ridge National Laboratory, Garlinburg, Tennessee, April 4-6, 1955.
2 Lacan 1989, 10.
3 Lacan 1986, 122.
4 Judson 1979, 275. Jacob renders the dictum in the following form: "Al Hershey, one of the most brilliant American specialists on bacteriophage, said that, for a biologist, happiness consists in working up a very complex experiment and then repeating it every day, modifying only one detail" (Jacob 1988, 236).
5 Jacob 1988, 234.
6 Heidegger 1977b, 118. "Die Zeit des Weltbildes" might be more appropriately translated as "The Epoch of Planetary Configuration."
7 Holmes 1985, xvi
8 Goethe 1988. A more accurate translation of "Der Versuch als Vermittler von Objekt und Subjekt" would be "The *Assay* as Mediator *of* Object and Subject" (emphasis added).
9 Kinder 1990.
10 Goethe 1988, 16.
11 Staiger 1966, letter of Goethe to Schiller, January 10, 1798.
12 Ibid., letter of Schiller to Goethe, January 12, 1798, 539-42.
13 Popper 1968, 107
14 Collins 1985.
15 Fleck 1979, 96, emphasis added. Ludwik Fleck, mainly because of his notion of "Denkstil," has often been misconstrued as a forerunner of the Kuhnain way of thinking in "paradigms."
16 Fleck 1979, 86.
17 Kuhn 1992, 14.
18 Jacob 1988, 9.
19 Lenoir 1992. See also Lenoir 1988, where a number of related positions are discussed.
20 Bachelard 1984, 139.
21 Ibid., chapter 6. See also Bachelard 1968.
22 As far as the history of molecular biology is concerned, one of the most brilliant examples is Jacob 1988. In recent years, we have witnessed a rapidly accumulating body of autobiographies from molecular biologists of the first generation. See, among others, Watson 1968, Luria 1984, McCarty 1985, Crick 1988, Kornberg 1989, Hoagland 1990. For a review of some of these works, see Abir-Arn 1991.
23 For a more fine-grained analysis, see Hentschel 1995.
24 Serres 1987, 191.
25 Latour 1987, 87-88.
26 For the notion of "constraint," see Galison 1995.
27 Sanger, Nicklen, and Coulson 1977.
28 Latour 1987, 131 and elsewhere.
29 Fleck 1979, 86, emphasis in first sentence omitted.
30 For more on scientific texts in biology, see Myers 1990; see also Bazerrnan 1988.

Bruno Latour

6 WHICH PROTOCOL FOR THE NEW COLLECTIVE EXPERIMENTS?[1]

Ladies and Gentlemen, we are all familiar with the notion of rules of methods for scientific experiments. Since the time of Bacon and Descartes, there is hardly a famous scientist who has not written down a set of rules to direct one's mind or, nowadays, to enhance the creativity of one's own laboratory, to organise one's discipline, or promote a new science policy. Even though these rules might not be enough to certify that interesting results will be obtained, they have been found useful nonetheless in establishing the state of the art. Equipped with those rules it is possible, according to their promoters, to say why some argument, behaviour, discipline, or colleague is or is not scientific enough.

Now the question before us tonight is certainly not to propose yet another set of rules to determine what is a scientific experiment or to offer advice on how to become even more scientific. For this task, anyway, I would be wholly incompetent. What I have chosen to explore with you is a rather new question which has only recently come to the foreground of public consciousness: namely, collective experiments. What are those collective or socio-technical experiments? Are they run in a totally wild manner with no rules at all? Would it be desirable to find rules to conduct them? What does it mean for the ancient definition of rationality and rational conduct? And, I will add, what does it mean for a specifically European conception of democracy? Such are the questions that, with your permission, I intend, not to try to solve but to touch upon tonight.

Laboratories Inside Out

That we are all engaged into a set of collective experiments that have spilled over the strict confines of the laboratories does not need more proof than the reading of the newspapers or the watching of the night TV news. At the time when I speak, thousands of officials, policemen, veterinarians, farmers, custom officers, firemen, are fighting all over Europe, indeed now all over the world, against the foot and mouth virus that is devastating so many countrysides. Nothing new in this, of course, since public health has been invented two centuries ago to prevent the spread of infectious diseases through quarantine and, later, disinfecting and vaccination. What is new, what is troubling, what requires our attention is that the present epizooty is due precisely to the collective decision not to vaccinate the animals. In this crisis, we are not faced, like our predecessors, with a deadly disease that we should fight with the weapons concocted inside the laboratory of Robert Koch or Louis Pasteur and their descendants: we find ourselves entangled in the unwanted—but wholly predictable—consequences of a decision to experiment, at the scale of Europe, on how long non-vaccinated livestock could survive without a new bout of this deadly disease. A nice case of what Ulrich Beck (1992) has called *manufactured risks*.

By mentioning this case, I am not trying to make you indignant; I am not claiming that *naturally* we should have vaccinated livestock; I am not saying it is a scandal because economic interests have taken precedence over public health and the welfare of farmers. There exist, I am well aware, many good reasons for the decision not to vaccinate. My point is different: a collective experiment has been tried out where farmers, consumers, cows, sheep, pigs, veterinarians, virologists have been engaged together. Has it been a well designed or a badly designed experiment? That is the question I want to raise.

In the time past, when a scientist or a philosopher of science was thinking of writing down rules of method, he (more rarely she) was thinking of a closed site, the laboratory, where a small group of specialised experts were scaling down (or scaling up) phenomena which they could repeat at will through simulations or modelling, before presenting, much later, their results, which could then, and only then, be scaled up, diffused, applied, or tried out. We recognise here the *trickling down* theory of scientific influence: From a confined centre of rational enlightenment, knowledge would emerge and then slowly diffuse out to the rest of society. The public could chose to learn the results of the laboratory sciences or remain indifferent to them, but it could certainly not add to them, dispute them, and even less contribute to their elaboration. Science was what was made inside the walls where white coats were at work. Experiments were undergone by animals, materials, figures and software. Outside the laboratory, borders began the realm of mere experience—not experiment (Dear 1990, 1995; Licoppe 1996).

It would be an understatement to say that nothing, absolutely nothing, has been left of this picture, of this trickling down model of scientific production.

First, the laboratory has extended its walls to the whole planet. Instruments are everywhere. Houses, factories, hospitals have become the subsidiaries of the labs. Think, for instance, of the global positioning system: thanks to this satellite network geologists and naturalists can now take measurements with the same range of precision outside and inside their laboratories. Think of the new requirements for tracability which are as stringent outside as those for inside the production sites. The difference between natural history, outdoor science, and lab science has slowly been eroded.

Second, it is well known from the development, for example, of patient organisations that many more people are formulating research questions, insisting on research agendas, than those who hold a PhD or wear a white coat. My colleague, Michel Callon, has been following for several years now a patient organisation in France, the AFM, which fights against orphan genetic diseases: they have not waited for results of molecular biology to trickle down to patients in wheel chairs: they have raised the money, hired the researchers, pushed for controversial avenues like genetic therapy, fired researchers, built an industry and in so doing they have been producing at once a new social identity and a new research agenda (Callon and Rabeharisoa 1999). The same can be said of many other groups, the best example being provided by the AIDS activists so well analysed by Steven Epstein (1996). And you would find the same situation throughout the whole ecological activism: If a crucial part of doing science is formulating the questions to be solved, it is clear that scientists are not alone in this. If in doubt on this point, ask the anti-nuclear militants about what type of research on energy they think laboratory scientists should be doing.

Third, the question of scale. Experiments are now happening at scale one and in real time, as it has become clear with the key question of global warming. To be sure, many simulations are being run; complex models are being tried out on huge computers, but the real experiment is happening on us, with us, through the action of each of us, on all of us, with all the oceans, high atmosphere and even the Gulf Stream—as some oceanographers argue (Broecker 1997)—participating in it. The only way to know if global warming is indeed due to anthropic activity is to try out and stop our noxious emissions to see then later, and collectively, what has happened. This is indeed an experiment but at scale one in which we are all embarked.

But then, what is now the difference with what used to be called a political situation: namely, what interests everyone and concerns everyone? None. That's precisely the point. The sharp distinction between scientific laboratories experimenting on theories and phenomena inside, and a political outside where non-experts were getting by with human values, opinions and passions, is simply evaporating under our eyes. We are now all embarked in the same collective experiments mixing humans and non-humans together—and no one is in charge. Those experiments made on us, by us, for us have no protocol. No one is given explicitly the responsibility of monitoring them. This is why a new definition of sovereignty is being called for.

When I am saying that the distinction between the inside and the outside of the laboratory has disappeared, I am not saying that from now on *all is political*. I am simply reminding you that contemporary scientific controversies are designing what Arie Rip and Michel Callon have called *hybrid forums* (Callon and Rip 1991). We used to have two types of representations and two types of forums: one that was in charge of representing things of nature—and here the word *representation* means accuracy, precision and reference—and another one which was in charge of representing people in society—and here the word *representation* meant faithfulness, election, obedience. One simple way to characterise our times is to say that the two meanings of representation have now merged into one around the key notion of spokesperson.

For instance, the global warming controversy is just one of those many new hybrid forums: some of those spokespersons represent the high atmosphere, others the lobbies of oil and gas, still others non-governmental organisations, still others represent, in the classical sense, their electors (with President Bush able to represent simultaneously his electors and the energy lobbies who have bought him up!). The sharp difference that seemed so important between those who represented things and those who represented people has simply vanished. What counts is that all those spokespersons are in the same room, engaged in the same collective experiment, talking at once about imbroglios of people and things. It does not mean that everything is political, but that a new politics certainly has to be devised, as Peter Sloterdijk has so forcefully argued in his vertiginous text "Regeln für den Menschenpark" (Sloterdijk 2000).

As I am sure you all know, the old word for *thing* does not mean what is outside the human realm, but a case, a controversy, a cause to be collectively decided in the *Thing*, the ancient word for assembly or forum in Old Icelandic as well as in Old German. Well, one can say, that things have become *things* again: "ein Ding ist ein Thing" (Thomas 1980). Have a look at the scientific as well as in the lay press, there is hardly a thing, a state of affair, which is not also, through litigation, protestation, a case; *une affaire* as we would say in French, *res* in Latin, *aitia* in Greek. Hence the expression I have chosen for this new politic: how to assemble the Parliament of Things (Latour 1993). Rules of method have become now rules, not to manage the Human Park, but to elaborate together the protocol of those collective experiments.

So what does the new division of labour look like? If I was not ending this lecture with some indication of the new configuration, you would be entitled, ladies and gentlemen, to say that I remain in the critical mood, unable to produce any positive version adjusted to the new situation.

In their new book soon to appear Michel Callon, Pierre Lascoumes, and Yannick Barthe, propose to replace the defunct notion of expert by the wider notion of co-researchers. As I have said at the beginning, we are all engaged, at one title or another, into the collective experiments on matters as different as climate, food, landscape, health, urban design, technical communication, and so on. As consumers, militants, citizens, we are all now co-researchers. There is a difference, to be sure, between all of us, but not the difference between knowledge producers and those who are bombarded by their applications. The idea of an *impact* of science and technology *on society* has been shipwrecked exactly as much as the weak notion of a *participation of the citizens into technology*. Now we have been made (most of the time unwillingly) all co-researchers and we are all led to formulate research problems, those who are *confined* in their laboratories as well as those that Callon and his colleagues call *outdoor* researchers, that is all of us. In other words, science policy, which used to be a specialised bureaucratic domain interesting a few hundreds of people, has now become an essential right of the new citizenry. The sovereignty over research agendas is much too important to be left to the specialists, especially when it is not in the hands of the scientists either, but in those of industry that no one has elected and that no one controls. Yes, we might be willing to participate in the collective experiments, but on the condition that we give our informed consent. Don't play on us any more the dirty tricks of considering all of us as the mere domain of applications of innovations concocted elsewhere. Look at what happened to those who believed genetically modified organisms could be made to *impact* European countryside. It does not mean people believe it is dangerous, nor does it mean that GMO are not safe—they might, as far as I am concerned, be totally safe and even indispensable for third world countries. But the question is not there anymore, as if we should accept anything as long as it is innocuous: the question has become again that of will and sovereignty: do we wish to live in this world? do we wish to draw that cosmogram? And if experts and modernists reply that there is one world only and that we have no choice to live in it or not, then let them say as well that there is no politics any more. When there is no choice of alternative, there is no Sovereign. What was true of the nation states, is becoming truer every day, under our very eyes, of our conflicting cosmos.

As I have argued at length in a book soon to appear in German, *Politics of Nature*, all of the rules of method for the collective experiment can be summarised by taking up again this magnificent slogan that our forefathers have chanted and chanted again, in building, through so many revolutions, their representative democracy: "No taxation without representation". Except that now, for the new technical democracies to be invented, it should read: "No innovation without representation". In the same way as the benevolent monarchies of the past imagined that they could tax us for our own good without us having a say on their budget because they alone were enlightened enough, in the same way, the new enlightened elite have been telling us for too long that there is only one best way for the innovation they have devised, and that we should simply follow them for our own good. Well, we might not be as enlightened as they are, but if the first Parliaments of the emerging nation-states were built to vote on budgets, the new Parliament of things have to be constructed to represent us so that we have a say on the innovations and decide for ourselves what is good for us. "No innovation, without representation".

A European Task?

Ladies and gentlemen, I want to bring this long and may be too hesitant lecture to a close, by offering a last proposition that has to do, this time, with Europe and its identity. As you are all too painfully aware, there seems no clear idea of what is specific to our sub-continent in those times of so called *globalisation*. I have always found this uneasiness pretty puzzling, since Europe, it is fair to say, has invented and developed in many ways the modernist regime of scientific and technical innovations—others of course had developed many sciences and techniques but never did they engage in the mad experiment of building their politics with them as well. But Europe is also a real life experiment, at an incredible scale, in multiculturalism, multinationalism, and in spite of that, it is trying to see how a common good can be slowly and carefully built. Nowhere else have so many fighting nation-states existed, so many provinces, regions, dialects, folklores and cultures. Nowhere else have world wars been waged to the bitter and deadly end. And yet, nowhere else have so many people engaged simultaneously into the cosmopolitic task—in the ordinary sense of the word—of living side by side in the same shared space, with the same Parliament, soon the same currency, and the same sense of democracy.

Now, I am asking you, why what is true of multiculturalism would not be true of multinaturalism as well. After all, if we have invented modernism, who else is better placed to, so to speak, disinvent modernity? No one else would do it, certainly not the United States which are too powerful, too sure of themselves, too deeply steeped in the modernity they have inherited without paying the costs—since others are bearing the cost for them (Todd 2002). Certainly not the many cultures who dream only, from Africa to the shores of Asia and Latin America, of being at last fully, utterly, and completely modernised—no wonder, alas, they took up at our own words! No, its Europe's chance, Europe's duty, Europe's responsibility to tackle first the perilous project of adding technical democracy to its old and venerable tradition of representative democracy. If we, Europeans, have learned the hard way how difficult it is to build a common good out of so many warring nation-states, we have a unique competence to learn, the hard way also, how to build a common world out of competing cosmos. Ladies and gentlemen, only those who have invented the premature unification of the whole world under the aegis of an imperialist nature, are well placed, now that nature has ended, to finally pay the price of the progressive, precautionous, modest, slow composition of the common world, this new name for politics. Thank you very much for your attention.

from: Schmidgen, Henning, Geimer, Peter & Dierig, Sven, ed. (2004), Kultur im Experiment, Kulturverlag Kadmos Berlin

REFERENCES

Beck, Ulrich (1992) *Risk Society. Towards a New Modernity.* London: Sage.

Beck, U. and Giddens, A. et al. (1994) *Reflexive Modernization. Politics, Tradition and Aesthetics in the Modern Social Order.* Stanford University Press.

Broecker, W.S. (1997) "Thermohaline Circulation, the Achilles Heel of Our Climate System", *Science* 278: 1582-1588, 28 November.

Callon, M.; Lascoumes, P. et al. (2001) *De la démocratie technique.* Paris: Le Seouil.

Callon, M. and Rabeharisoa, V. (1999) *Le pouvoir des malades.* Presses de l'Ecole nationale des mines de Paris.

Callon, M. and Rip, A. (1991) "Forums hybrides et négociations des normes socio-techniques dans le domaine de l'environnement". *Environnement, Science et Politique. Cahiers du Germes* Paris. N. 13: 227-228, Groupe d'Exploitation et de Recherches Multidisciplinaires sur l'Environnement et la Société.

Dear, Peter (1990) *Experiment as Metaphor In the Seventeenth Century.* 1–26.

Dear, Peter (1995) *Discipline and Experience: The Mathematical Way in the Scientific Revolution* Chicago: University of Chicago Press.

Dewey, J. (1927) *The Public and Its Problems.* Athens: Ohio University Press. (1954 edition by Chicago: Swallow Press.)

Dratwa, Jim (2003) *Taking Risks with the Precautionary Principle* PhD Thesis

Epstein, S. (1996) *Impure Science. Aids, Activism and the Politics of Knowledge.* Berkeley: University of California Press.

Fleck, Ludwing (1935) *Genesis and Develpment of a Scientific Fact.* University of Chicago Press.

Fox–Keller, E. (2000) *The Century of the Gene.* Cambridge Massachusetts: Harvard University Press.

James, William (1907) *Pragmatism. A New Name for Some Old Ways of Thinking followed by The Meaning of Truth.* 1975 edition by Harvard University Press, Cambridge, Massachusetts.

Jonas, Hans (1984) *The Imperative of Responsibility.* University of Chicago Press.

Jurdant, B. (Ed.) (1998) *Impostures intellectuelles. Les malentendus de l'affaire Sokal.* Paris: La Découverte.

Kupiec, J.J. and Sonigo, P. (2000) *Ni Dieu ni gène.* Paris: Le Seuil, Collection Science Ouverte.

Latour, Bruno (1993) *We Have Never Been Modern.* Cambridge Massachusetts: Harvard University Press. (English version at Harvard University Press, spring 2004 translation by Cathy Porter).

Latour, Bruno (1999a) *Pandora's Hope. Essays on the reality of science studies.* Cambridge Massachusetts: Harvard University Press.

Latour, Bruno (1999b) *Politiques de la nature. Comment faire entrer les sciences en démocratie.* Paris: La Découverte.

Latour, Bruno and Weibel, Peter (Eds.) (2002) *Iconoclash. Beyond the Image Wars in Science, Religion and Art.* Cambridge Massachusetts: MIT Press.

Lewontin, Richard (2000) *The Triple Helix. Gene, Organism and Environment.* Cambridge Massachusetts: Harvard University Press.

Licoppe, Christian (1996) *La formation de la pratique scientifique. Le discours de l'expérience en France et en Angleterre (1630–1820).* Paris: La Découverte.

Lippmann, Walter (1922) *Public Opinion.* New York: Simon & Schuster.

Poovey, Mary (1999) *History of the Modern Fact. Problems of Knowledge in the Sciences of Wealth and Society.* Chicago University Press.

Rheinberger, Hans–Jorg (1997) *Toward a History of Epistemic Thing. Synthetizing Proteins in the Test Tube.* Stanford University Press.

Ryan, A. (1995) *John Dewey and the High Tide of American Liberalism.* New York: Norton l.

Sloterdijk, P. (2000) *Régles pour le parc humain.* Mille et une nuits, Paris.

Stengers, I. (1996) *Cosmopolitiques. Tome 1: La guerre des sciences* La découverte &Les Empêcheurs de penser en rond, Paris.
Thomas, Y. (1980) "Res, chose et patrimoine (note sur le rapport sujet – objet en droit romain)", *Archives de philosophie du droit.* t 25: 413–426.
Todd, Emmanuel (2002) *Après l'Empire, essai sur la décomposition du système amèricain.* Paris: Gallimard.
Tresch, J. (2001) *Mechanical Romanticism: Engineers of the Artificial Paradise.* PhD Thesis, Department of History and Philosophy of Science. University of Cambridge.

NOTES

1 "Regeln für die neuen wissenschaftlichen und sozialen Experimente" prepared for the Darmsdadt Colloquium plenary lecture, 30th March 2001. Original document avalaible in: http://www.ensmp.fr/~latour/poparticles/poparticle/p095.html. The printed version of this article is available in:
Schmindgen, Henning Peter Geimer, and Sven Dierig (eds.) (2004) *Kultur im Experiment* Kulturverlag Kadmos: Berlin, Juni.

John Law

7 SCIENTIFIC PRACTICES

> ... tools only exist in relation to the interminglings they make possible or that make them possible. The stirrup entails a new man-horse symbiosis that at the same time entails new weapons and new instruments. Tools are inseparable from symbioses or amalgamations defining a Nature-Society machinic assemblage. They presuppose a social machine that selects them and takes them into its 'phylum': a society is defined by its amalgamations, not by its tools. Similarly, the semiotic or collective aspect of an assemblage relates not to a productivity of language but to regimes of signs, to a machine of expression whose variables determine the usage of language elements. These elements do not stand on their own any more than tools do.
>
> (Deleuze and Guattari 1988, 90)

> A proposition, contrary to a statement, includes the world in a certain state. ... Thus a construction is not a representation from the mind or from the society about a thing, an object, a matter of fact, but the engagement of a certain type of world in a certain kind of collective.
>
> (Latour 1997, xiii-xiv)

Inscription Devices and Realities

In October 1975 a young French philosopher arrived at the Salk Institute in San Diego. Called Bruno Latour, he later wrote that his 'knowledge of science was non-existent; his mastery of English was very poor' (Latour and Woolgar 1986, 273). He watched the work of the Salk Institute endocrinologists for nearly two years and then wrote a book about it with sociologist of science Steve Woolgar. Called *Laboratory Life,* this appeared in 1979 and, with books by one or two others, helped to create a new field, that of the *ethnography of science.*

As we move through the present book we will look over the shoulders of ethnographers as they visit scientific laboratories, clinics, hospitals, religious ceremonies and managerial meetings. We will also watch the work of social scientists—and others—as they produce knowledge in practice. So what do ethnography of knowledge practices tell us? The answer is that ethnography lets us see the relative messiness of practice. It looks behind the official accounts of method (which are often clean and reassuring) to try to understand the often ragged ways in which knowledge is produced in research. Importantly, it doesn't necessarily distinguish very cleanly between science, medicine, social science, or any other versions of inquiry. Distinctions such as these tend to go out of focus in the welter of knowledge practices uncovered by ethnography. It also tends to find continuities between natural and social science. Physicists may have their instruments, but so too do sociologists. Much that we learn about the practice of natural science is also applicable to social science.

Thus the first take-home message from Latour and Woolgar is that what the authors called 'the tribe of scientists' (1986, 17) is not very different from any other tribe. Scientists have a culture. They have beliefs. They have practices. They work, they gossip, and they worry about the future. And, somehow or other, out of their work, their practices and their beliefs, they produce knowledge, scientific knowledge, accounts of reality. So how do they do this? How do they make knowledge?

The ethnographers of science are usually more or less *constructivist.* That is, they argue that scientific knowledge is constructed in scientific practices. This, it should be noted, is *not* at all the same thing as saying it is constructed by scientists. Thus we will see that practices include, and imply, instruments, architectures, texts—indeed a whole range of participants that extend far beyond people. But the process of building scientific knowledge is also an active matter. It takes work and effort. The argument is that it is wrong to imagine that nature somehow impresses its reality directly on those who study it if they just set aside their own biases. The picture of science offered by Merton is not right. But how is this construction done?

Different ethnographers respond to this question in somewhat different ways. However Latour and Woolgar, whom I follow here, explore it materially. They wouldn't call themselves 'materialists' because they do not think that everything derives from, or can be ultimately explained in, material terms. Nevertheless, they are very much into *materiality.* This means that they focus in the first instance on the physical stuff of the laboratory, and how this is laid out architecturally. For instance, it has a chemistry section, a physiology section, and then there is a location with desks and word processors which is mainly to do with paperwork. Then they talk about the way materials move around. Energy, money, chemicals, people, animals, instruments, tools, supplies, and papers of all kinds, move into the laboratory. At the same time, people and (different) papers and maybe instruments, together with debris and waste, move out. Looked at as a system of material production, then, the major product of the laboratory turns out to be *texts.* These are very expensive: at 1979 prices they cost about $30,000 each. No doubt the figure would be much higher now.

If the Salk Laboratory is a system of material production then how are its various material resources turned into texts? Latour and Woolgar trace this through a number of moves. Step one: they observe that 'the desk . . . appears to be the hub of our productive unit' (1986, 48). At the desk two kinds of texts are juxtaposed: on the one hand some come from outside the laboratory, such as scientific articles or books; on the other hand some originate from within the laboratory. But where do these come from? The answer is that they are produced by what they call *inscription devices.*

So this is the second step in their argument. An inscription device is a system (often including, though not reducible to, a machine) for producing inscriptions, or traces, out of materials that take other forms:

> an inscription device is any item of apparatus or particular configuration of such items which can transform a material substance into a figure or a diagram which is directly usable by one of the members of the office space.
>
> (1986, 51)

For instance, an inscription device might start out with rats. These would be sacrificed to produce extracts which would be placed in small test tubes. Then those test tubes would be placed in a machine, for instance a radiation detector, which would convert them into an array of figures or inscriptions on a sheet of paper. These inscriptions would be said—or assumed—to have a direct relation to 'the original substance'.

At this point, stage three, something interesting happens. Latour and Woolgar argue that *the process of producing the traces melts into the background:*

> The final diagram or curve thus provides the focus of discussion about properties of the substance. The intervening material activity and all aspects of what is often a prolonged and costly process are bracketed off in discussions about what the figure means.
>
> (1986, 51)

The argument is thus that *the materiality of the process gets deleted.* (Perhaps this is why 'constructivism' is often mistakenly thought to be about a purely human activity.) For what is subsequently manipulated is not the rats themselves. It is not even the extracts from the rats. Rather it is curves derived from figures from the relevant inscription devices. It is the curves that get juxtaposed with one another on the desks of the researchers.

The fourth step in the story is a process of isolating, detecting, and naming substances:

> Samples of brain extract are subjected to a series of *discriminations.* ... This involves the use of some stationary material (such as a gel or a piece of blotting paper) as a selective sift which delays the gradual movement of the sample of brain extract. ... As a result of this process, samples are transformed into a large number of fractions, each of which can be

scrutinised for physical properties of interest. The results are recorded in the form of several peaks on graph paper. Each of these peaks represents a discriminated fraction, one of which may correspond to [a] ... discrete chemical entity. ... In order to discover whether the entity is present, the fractions are taken back to the physiology section of the laboratory and again take part in an assay. By superimposing the result of this last assay with the result of the previous purification, it is possible to see an overlap between one peak and another. If the overlap can be repeated, the chemical fraction is referred to as a 'substance' and is given a name.

(1986, 60)

This is very important. Latour and Woolgar are telling us that it is *more or less stable similarities between curves* that allow the scientists to say that they have isolated a 'substance'. It is the relative similarities of successive curves that allow the laboratory workers to name a 'substance'. By contrast, 'elusive and transitory' substances—witnessed by curves that appear and disappear—come to be known as 'artefacts' and are disregarded.

Though some of their language is unusual, and, yes, they have taken us away from empiricism, perhaps what Latour and Woolgar have told us so far is not too surprising. But with the next step we move towards the unexpected:

The central importance of this material arrangement [of laboratory inscription devices] is that none of the phenomena 'about which' participants talk could exist without it. Without a bioassay, for example, a substance could not be said to exist. The bioassay is not merely a means of obtaining some independently given entity; the bioassay constitutes the construction of the substance.

(1986, 64)

'Without a bioassay, for example, a substance could not be said to exist.' And this is not simply a way of speaking. Here they are again:

It is not simply that phenomena *depend on* certain material instrumentation; rather, the phenomena *are thoroughly constituted by* the material setting of the laboratory. The artificial reality, which participants describe in terms of an objective entity, has in fact been constructed by the use of inscription devices.

(1986, 64)

This, then, is their fifth point. It is that *particular realities are constructed by particular inscription devices* and practices. Let me emphasise that: *realities* are being *constructed.* Not by people. But in the practices made possible by networks of elements that make up the inscription device—and the networks of elements within which that inscription device resides. The realities, they are saying, simply don't exist without their matching inscription devices. And, implicitly at least, they are also saying that such inscription devices—and even more so their particular products—are elaborate and networked arrangements that are more or less uncertain, more or less able to hold together, and more or less precarious. [...]

In sum, Latour and Woolgar take us some distance from everyday Euro-American expectations about out-thereness. Reality is neither independent nor anterior to its *apparatus* of production. Neither is it definite and singular until that *apparatus* of production is in place. Realities are made. They are *effects of the apparatuses of inscription.* At the same time, since there are such apparatuses *already* in place, we also live in and experience a real world filled with real and more or less stable objects.

A Routinised Hinterland: Making and Unmaking Definite Realities

So why is scientific reality relatively stable, at least a lot of the time? Latour and Woolgar suggest that we might think about this in terms of *cost.* The argument is that undermining the relations embedded in received statements is expensive:

> the set of statements considered too costly to modify constitute what is referred to as reality. Scientific activity is not 'about nature,' it is a fierce fight to *construct* reality. The *laboratory* is the workplace and the set of productive forces, which makes construction possible. Every time a statement stabilises, it is reintroduced into the laboratory (in the guise of a machine, inscription device, skill, routine, prejudice, deduction, programme, and so on), and it is used to increase the difference between statements. The cost of challenging the reified statement is impossibly high. Reality is secreted.
>
> (1986, 243)

'Reality is secreted.' Notice that this posits a kind of feedback loop. Statements stabilise, and then recycle themselves back into the laboratory. This means that once they are demodalised, *yesterday's modalities become tomorrow's hinterland.* And, as a part of this they tend to change in their material form:

> The mass spectrometer is the reified part of a whole field of physics; it is an actual piece of furniture which incorporates the majority of an earlier body of scientific activity.
>
> (1986, 242)

So why and how do they change their material form? A part of the answer is that it is easier to produce statements about realities—easier to produce realities—when these take standardised and transportable forms. Latour and Woolgar talk of reification, but perhaps the notion of *routinisation* better draws attention to what is most important. We saw above that the practice of fitting bits and pieces together to produce more or less stable traces is a precarious business. Much goes wrong in laboratory science. But if machines and skills and statements can be turned into packages, then so long as everything works (this is always uncertain) there is no longer any need to individually assemble all the elements that make up the package, and deal with all the complexities. It is like buying a personal computer rather than understanding the electronics, and the physics embedded in the electronics and assembling one out of components. Thus in the above example the field of physics that is the hinterland of the mass spectrometer can be taken for granted. It does not have to be rebuilt or even understood by those who use the instrument. One sociology of science literature talks of 'standardised packages'. This is the point: in this way of thinking all the reality-describing and reality-making of natural (and social) science practices surfs on more or less provisional standardised packaes that are, form part of, or support, inscription devices and practices. At the beginning of this chapter I cited Latour:

> A proposition, contrary to a statement, includes the world in a certain state . ... Thus a construction is not a representation from the mind or from the society about a thing, an object, a matter of fact, but the engagement of a certain type of world in a certain kind of collective.
>
> (Latour 1997, xiii—xiv)

Latour, here twenty years on, is talking about Isabelle Stengers's philosophy of science (and his talk of propositions rather than statements is a small but potentially misleading change in vocabulary). But the overall argument remains the same. It is not a matter of words representing things. Words and worlds go together. Propositions (as he is now calling them) include realities—include a collective. Include and grow from what I am calling the hinterland.

Certain additional consequences follow. The hinterland produces specific more or less routinised realities and statements about those realities. But this implies that countless other realities are being *un-made* at the same time—or were never made at all. To talk of 'choices' about which realities to make is too simple and voluntaristic. The hinterland of standardised packages at the very least shapes our 'choices'. We who 'choose' embody and carry a bundle of hinterlands. Nevertheless there are a whole lot of realities that are not, so to speak, real, that would indeed have been so if the apparatus of reality-production had been very slightly different.

A further and related implication is that the hinterland produces certain *classes* of realities and reality-statements—but not others. Some kinds of standardised inscription devices and practices are current. Some classes of reality are more or less easily producible. Others, however, are not—or were never cobbled together in the first place. So the hinterland also defines an overall geography—a topography of reality-possibilities. Some classes of possibilities are made thinkable and real. Some are made less thinkable and less real. And yet others are rendered completely unthinkable and completely unreal.

The economic metaphor suggests that it is easier and cheaper to create new inscription devices, new statements and new realities by building on to the routinised black boxes that are already available. It also suggests that as the process goes along it becomes more and more difficult and expensive to ignore or to undo the routines and create others and alternative realities. [...]

Here are a few thoughts. First, even though its argument is unfamiliar, it is plausible. Even if it doesn't fit the standard Euro-American justifications, Latour and Woolgar's account fits the *practices* of natural and social science. The findings of their ethnography are neither empirically weird nor theoretically strained. They explain perfectly well why scientists (and social scientists and lay people) tend to be committed to a strong version of out-thereness. But at the same time they also show how this is consistent with the idea that out-thereness is something enacted in practice. As I have shown above, scientists are caught up in a hinterland that has both been created and yet is relatively obdurate because it is too difficult to overturn.

Latour and Woolgar's argument applies just as well to our social sciences. We too have our instruments of research. We too reflect on and work within the obdurate realities produced by the hinterland of those instruments. For instance, statistics do not exist *sui generis*. As is obvious, they have to be created. Indeed there has been considerable historical work on the way in which this has been achieved over a couple of centuries or more through the medium of elaborate systems of tallying, measuring and quantifying in such forms as censuses, timekeeping (or time-making), surveying and economic data- creation. Such apparatuses, the hinterlands of much of social science, embed and enact many assumptions about the nature of the social. Arguably, 'the social' was brought into being in these apparatuses, as they developed and carried strategies of social and state control. By now however, with so many daily practices (public and private) dependent on official and other statistics, their reversibility is in doubt. It is possible to tinker with them—but overall, undoing them would be extremely expensive both literally and metaphorically. The result is both that we have come to live, and are made, in a social reality that is partly quantitative in quite specific ways, and that much of this hinterland is bundled into and constitutive of social science research. We might add that parts of it have also been produced by social science.

None of this is to say that these statistics are wrong. They may be criticised for this or that particular failing, but this is not the point. Rather they and the relations in which they are located are hinterlands and social realities out-there that both enable and constrain any work in social science. They set limits to the conditions of social science possibility. Overall, then, this is the first reason for taking the arguments of Latour and Woolgar seriously. Though their argument about enacted realities sounds counter-intuitive, it is consistent with our Euro-American intuitions that realities, natural and social, are pretty solid. To say that something has been 'constructed' along the way is not to deny that it is real.

Second, and just as important, their argument helps us to think differently and more creatively about method. In particular, the suggestion that specific forms of out-thereness are enacted and re-enacted makes it possible to think about which realities it might be best to bring into being. This, as I hope I have made clear, is not a simple or trivial question of choosing the version of out-thereness that happens to suit. 'Choice', if this is an appropriate term at all, is limited by the need to relate to and build appropriate hinterlands that will sustain statements about reality. Philosopher Isabelle Stengers puts the argument in slightly different terms:

> no scientific proposition describing scientific activity can, in any relevant sense, be called 'true' *if it bas not attracted 'interest'*. To interest someone does not necessarily mean to gratify someone's desire for power, money or fame. Neither does it mean entering into preexisting interests. To interest someone in something means, first and above all, to act in such a way that this thing—apparatus, argument, or hypothesis ... —can concern the person, intervene in his or her life, and eventually transform it. An interested scientist will ask the question: can I incorporate this 'thing' into my research?
>
> (Stengers 1997, 82-83)

So this is not a trivial matter. 'Interesting' is not necessarily easy. Nevertheless, the implications are profound. If out-therenesses are constructed or enacted rather than sitting out there waiting to be discovered, then it follows that their truth or otherwise is only one of the criteria relevant to their creation. Politics in one form or another also becomes important. But the moment we acknowledge this we are faced with new questions. What kind of out-therenesses are possible? Which are so embedded that they cannot be undone? Where might we try to undo or redo them? How might we try to nudge research programmes in one direction rather than another? To bend a phrase, if we think in this way then reality is no longer destiny. [...]

When I want to refer to method in this extended manner I will usually speak of *method assemblage.* I will return to and redefine this term several times in what follows, and especially in Chapters 3 and 5. However I will start by noting that the term 'assemblage' comes from the English translation of Deleuze and Guattari's *Mille Plateaux* (see the citation that begins this chapter). Helen Verran and David Turnbull say that for these authors an assemblage:

> is like an episteme with technologies added but that connotes the ad hoc contingency of a collage in its capacity to embrace a wide variety of incompatible components. It also has the virtue of connoting active and evolving practices rather than a passive and static structure.
>
> (Watson-Verran and Turnbull 1995, 117)

Here Verran and Turnbull have caught exactly what is needed. An *assemblage* (without the method) is an episteme plus technologies. It is ad hoc, not necessarily very coherent, and it is also active.

In Deleuze and Guattari the English term 'assemblage' has been used to translate the French 'agencement'. Like 'assemblage', 'agencement' is an abstract noun. It is the action (or the result of the action) of the verb 'agencer'. In French 'agencer' has a wide range of meanings. A small French—English dictionary tells us that it is 'to arrange, to dispose, to fit up, to combine, to order'. A large French dictionary offers dozens of synonyms for 'agencement' which together reveal that the term has no single equivalent in English. This means that while 'assemblage' is not exactly a mistranslation of 'agencement' much has got lost along the way. In particular the notion has come to sound more definite, clear, fixed, planned and rationally centred than in French. It has also come to sound more like a state of affairs or an arrangement rather than an uncertain and unfolding process. If 'assemblage' is to do the work that is needed then it needs to be understood as a tentative and hesitant unfolding, that is at most only very partially under any form of deliberate control. It needs to be understood as a verb as well as a noun. Here is Derrida (of course in translation):

> . . . the word sheaf seems to mark more appropriately that the assemblage to be proposed as the complex structure of a weaving, an interlacing which permits the different threads and different lines of meaning—or of force—to go off again in different directions, just as it is always ready to tie itself up with others.
>
> (Derrida 1982, 3)

Note that. A *'complex structure of a weaving'*. A *'sheaf'*. And here are Deleuze and Claire Parnet:

In a multiplicity, what counts are not the terms or the elements, but what there is 'between', the between, a set of relations which are not separable from each other.

(Deleuze and Parnet 1987, viii)

So assemblage is a process of bundling, of assembling, or better of recursive self-assembling in which the elements put together are not fixed in shape, do not belong to a larger pre-given list but are constructed at least in part as they are entangled together. This means that there can be no fixed formula or general rules for determining good and bad bundles, and that (what I will now call) 'method assemblage' grows out of but also *creates* its hinterlands which shift in shape as well as being largely tacit, unclear and impure. [...]

from: Law, John (2004): *After Method: Mess in Social Science Research*, London, © Taylor & Francis Books UK

Julian Bleecker

8 DESIGN FICTION

[...]

From this starting place, I think of design as a kind of creative, imaginative authoring practice—a way of describing and materializing ideas that are still looking for the right place to live. A designed object can connect an idea to its expression as a made, crafted, instantiated object. These are like props or conversation pieces that help speculate, reflect and imagine, even without words. They are things around which discussions happen, even with only one other person, and that help us to imagine other kinds of worlds and experiences. These are material objects that have a form, certainly. But they become real before themselves, because they could never exist outside of an imagined use context, however mundane or vernacular that imagined context of social practices might be. Designed objects tell stories, even by themselves.

If design can be a way of creating material objects that help tell a story what kind of stories would it tell and in what style or genre? Might it be a kind of half-way between fact and fiction? Telling stories that appear real and legible, yet that are also speculating and extrapolating, or offering some sort of reflection on how things are, and how they might become something else?

Design fiction as I am discussing it here is a conflation of design, science fact, and science fiction. It is a amalgamation of practices that together bends the expectations as to what each does on its own and ties them together into something new. It is a way of materializing ideas and speculations without the pragmatic curtailing that often happens when dead weights are fastened to the imagination.

The notion that fiction and fact could come together in a productive, creative way came up a couple of years ago while participating in a reading group where a colleague presented a draft of a paper that considered the science fiction basis of the science fact work he does. He saw a relationship between the creative science fiction of early television in Britain and the shared imaginary within the science fact world of his professional life. There were linkages certainly, suggesting that science fiction and science fact can share common themes, objectives and visions of future worlds.

My colleague was not saying that the science of fact and the science of fiction were the same. In fact, he was explicitly not conflating the two. Nevertheless, coming from a computer science professor I found this idea intriguing in itself. It was certainly something to mull over.[1] What was percolating in my mind was this liminal possibility of a different approach to doing the same old tired stuff. This notion presented a new tact for creative exploration—a different approach to doing research.

I wondered—rather than an approach that adheres dogmatically to the principles of one discipline, where anything outside of that one field of practice is a contaminant that goes against sanctioned ways of working, why not take the route through the knotty, undisciplined tangle? Why not employ science fiction to stretch the imagination?

Throw out the disciplinary constraints one assumes under the regime of fact and explore possible fictional logics and assumptions in order to reconsider the present.

Finally, I recognized that the science fact and the science fiction he was discussing were quite closely related *in practice* and probably quite inextricably and intimately tangled together, more so then the essay may have been letting on. In other words, I began to wonder if science fact and science fiction are actually two approaches to accomplishing the same goal—two ways of materializing ideas and the imagination.

My bias—arrived at through a mix of skepticism, experience, and desire to do things differently—is that, generally, it seems that science fiction does a much better job, if only in terms of its capacity to engage a wider audience which oftentimes matters more than the brilliant idea done alone in a basement.

My question is this—*how can science fiction be a purposeful, deliberate, direct participant in the practices of science fact?*

This is what this essay on design fiction is about. It is one measure manifesto, one measure getting some thinking off my chest, one measure reflection on what I think I have been doing all along, and one measure explanation of why I am doing what I am doing.

Science fiction can be understood as a kind of writing that, in its stories, creates prototypes of other worlds, other experiences, other contexts for life based on the creative insights of the author. Designed objects—or designed fictions—can be understood similarly. They are assemblages of various sorts, part story, part material, part idea-articulating prop, part functional software. The assembled design fictions are component parts for different kinds of near future worlds. They are like artifacts brought back from those worlds in order to be examined, studied over. They are puzzles of a sort. A kind of object that has lots to say, but it is up to us to consider their meanings. They are complete specimens, but foreign in the sense that they represent a corner of some speculative world where things are different from how we might imagine the "future" to be, or how we imagine some other corner of the future to be. These worlds are "worlds" not because they contain everything, but because they contain enough to encourage our imaginations, which, as it turns out, are much better at filling out the questions, activities, logics, culture, interactions and practices of the imaginary worlds in which such a designed object might exist. They are like conversations pieces, as much as a good science fiction film or novel can be a thing with ideas embedded in it around which conversations occur, at least in the best of cases. A design fiction practice creates these conversation pieces, with the conversations being stories about the kinds of experiences and social rituals that might surround the designed object. Design fiction objects are totems through which a larger story can be told, or imagined or expressed. They are like artifacts from someplace else, telling stories about other worlds.

What are these stories? They are whatever stories you want to tell. They are objects that provide another way of expressing what you're thinking, perhaps before you've even figured out what you imagination and your ideas mean. Language is a tricky thing, often lacking the precision you'd like, which is why conversation pieces designed to provoke the imagination, open a discussion up to explore possibilities and provoke new considerations that words by themselves are not able to express. Heady stuff, but even in the simplest, vernacular contexts, such stories are starting points for creative exploration.

Design is the materialization of ideas shaped by points-of-view and principles that tell you "how" to go about materializing an idea. Principles are like specifications of a sort, only the kind I am describing are of a more interpretive, imaginative and elastic sort. Not like engineering specifications, or the typical list of contents one finds in most any designed object—especially gadgets, like the flavors of WiFi, types of USB, quantities of gigabytes, diagonal screen inches, etc. Design principles are like the embedded DNA of a design, but can be as much a DNA about experiences to be had as instrumental measurements and adherence to manufacturing codes and trademark badges.

Design fiction is a way of exploring different approaches to making things, probing the material conclusions of your imagination, removing the usual constraints when designing for massive market commercialization—the ones that people in blue shirts and yellow ties call "realistic." This is a different genre of design. Not realism, but a genre that is forward looking, beyond incremental and makes an effort to explore new kinds of social interaction rituals. As much as science fact tells you what is and is not possible, design fiction understands constraints differently. Design fiction is about creative provocation, raising questions, innovation, and exploration.

Environment matters for these unconventional approaches. I play in a studio that's really exceptional, with incredibly creative designers whose have excellent listening skills and do not start with assumptions that are euphemisms for constraints and boundaries and limits. I'm not just saying that, its a point of pride in the studio. We don't design products, if such is taken to mean the product of manufacturing plants, rather than the product of active, thoughtful imaginations. But we do design provocations that confront the assumptions about products, broadly. Our provocations are objects meant to produce new ways of thinking about the near future, optimistic futures, and critical, interrogative perspectives. We clarify and translate strategic vectors, using design to investigate the many imaginable near futures. It's a way of enhancing the corporate imagination, swerving conversations to new possibilities that are reasonable but often hidden in the gluttony of overburdened markets of sameness. Running counter to convention is part of what some kinds of science fiction—rather, design fiction—allows for. This is especially valuable in the belly of a large organization with lots of history and lots of convention.

Design fiction is a mix of science fact, design and science fiction. It is a kind of authoring practice that recombines the traditions of writing and story telling with the material crafting of objects. Through this combination, design fiction creates socialized objects that tell stories—things that participate in the creative process by encouraging the human imagination. The conclusion to the designed fiction are objects with stories. These are stories that speculate about new, different, distinctive social practices that assemble around and through these objects. Design fictions help tell stories that provoke and raise questions. Like props that help focus the imagination and speculate about possible near future worlds—whether profound change or simple, even mundane social practices.

Design fiction does all of the unique things that science-fiction can do as a reflective, written story telling practice. Like science fiction, design fiction creates imaginative conversations about possible future worlds. Like some forms of science fiction, it speculates about a near future tomorrow, extrapolating from today. In the speculation, design fiction casts a critical eye on current object forms and the interaction rituals they allow and disallow. The extrapolations allow for speculation without the usual constraints introduced when "hard decisions" are made by the program manager whose concerns introduce the least-comon denominator specifications that eliminate creative innovation. Design fiction is the cousin of science fiction. It is concerned more about exploring multiple potential futures rather than filling out the world with uninspired sameness. Design fiction creates opportunities for reflection as well as active making.

Design fiction works in the space between the arrogance of science fact, and the seriously playful imaginary of science fiction, making things that are both real and fake, but aware of the irony of the muddle—even claiming it as an advantage. It's a design practice, first of all—because it makes no authority claims on the world, has no special stake in canonical truth; because it can work comfortably with the vernacular and pragmatic; because it has as part of its vocabulary the word "people" (not "users") and all that implies; because it can operate with wit and paradox and a critical stance. It assumes nothing about the future, except that there can be simultaneous futures, and multiple futures, and simultaneous-multiple futures—even an end to everything.

In this way design fiction is a hybrid, hands-on practice that operates in a murky middle ground between ideas and their materialization, and between science fact and science fiction. It is a way of probing, sketching and exploring ideas. Through this practice, one bridges imagination and materialization by modeling, crafting things, telling stories through objects, which are now effectively conversation pieces in a very real sense. A bit like making science fact prototypes, or props for a science fiction film, but not quite. We'll get to the "how" later.

When I think of design this way, it feels like it should be understood slightly differently from the all-encompassing "design," which is why I am referring to it as "design fiction."

from: Bleecker, Julian (2009): *Design Fiction: A short essay on design, science, fact and fiction*, Near Future Laboratory

NOTES

1 It was also a bit of a reminder of some earlier work I had done while a graduate student, working on Virtual Reality at the University of Washington, Seattle where an informal rite was to thoroughly read William Gibson's *Neuromancer* and the Cyberpunk manifesto by Gibson and Bruce Sterling *Mirrorshades*. More on this later.

Anthony Dunne

9 THE ELECTRONIC AS POST-OPTIMAL OBJECT

> As new technical developments alter the object and make it "intelligent," they also set the object on a plane with no prior cultural references ... although the physical aspects of these objects are still within the world of materials, their operation and their very state of being is well beyond the manipulation of matter and has more to do with information exchange than with form.
>
> —E. MANZINI, *The Material of Invention*

Most designers of electronic objects have responded to this challenge by accepting a role as a semiotician, a companion of packaging designers and marketers, creating semiotic skins for incomprehensible technologies.

From Banham writing about portable radios in the 1970s, through the plethora of essays on "product semantics" in the 1980s, to Norman Bolz's 1992 essay "The Meaning of Surface," the treatment of the electronic object as a package for technology, designed to communicate use, cultural meaning, and corporate identity through its surface, has been thoroughly explored. The electronic object accordingly occupies a strange place in the world of material culture, closer to washing powder and cough mixture than to furniture and architecture, and is subject to the same linguistic discipline as all package design, that of the sign. It is lost somewhere between image and object, and its cultural identity is defined in relation to technological functionalism and semiotics.

This chapter considers three perspectives on the electronic object: "The Electronic as Lost Object" briefly discusses theoretical perspectives, "The Electronic as Object" focuses on design approaches, and "The Electronic as Post-optimal Object" introduces the idea of the "post-optimal" object.

The Electronic as Lost Object

A Technological Perspective

From a technological perspective the theories of Jean Baudrillard and Paul Virilio are a stimulating source of ideas about the effects of electronic technology on the way we experience and think about ourselves, objects, and environments. Their provocative fusions of analysis and imagery offer a rich inspiration while remaining grounded in reality. But there is a danger that if designers are seduced by this, their designs will become mere illustrations of descriptions of electronic objects. Designers of electronic objects are already familiar with the kinds of technologies analyzed by these writers. It is more important to extend the range of cultural values, building on what is already understood, rather than illustrating it.

Some writers on the social history of technology present the ideological dimension of everyday technologies, even if these are often pre-electronic. This is useful to critique the human factors "community," who have developed a view of the electronic object, derived from computer science and cognitive psychology, that is extremely influential in the computer industry; see, for example, Don Norman's (1988) *The Psychology of Everyday Things.*

A serious problem with the human factors approach though, in relation to this project, is its uncritical acceptance of what has been called by Bernard Waites (1989) the "American Ideology," or the ideological legitimation of technology:

> All problems whether of nature, human nature, or culture, are seen as "technical" problems capable of rational solution through the accumulation of objective knowledge, in the form of neutral or value-free observations and correlations, and the application of that knowledge in procedures arrived at by trial and error, the value of which is to be judged by how well they fulfil their appointed ends. These ends are ultimately linked with the maximisation of society's productivity and the most economic use of its resources, so that technology, in the American Ideology, becomes "instrumental rationality" incarnate, the tools of technocracy. (31)

The result, as the computer industry merges with other industries, is that the optimization of the psychological fit between people and electronic technology, for which the industry strives, is spreading beyond the work environment to areas such as the home that have so far acted as a counterpoint to the harsh functionality of the workplace. When used in the home to mediate social relations, the conceptual models of efficient communication embodied in office equipment leave little room for the nuances and quirks on which communication outside the workplace relies so heavily.

Writing on electronic art might seem a good source of ideas on the electronic object, but, surprisingly, electronic art has become so technology-driven that it seems concerned only with the aesthetic expression of technology for its own sake. Rather than relating the impact of technology to everyday life, art criticism in this area glamorizes technology as a source of aesthetic effect to be experienced in galleries. The exceptions tend to be based on electronic systems rather than objects (e.g., in the work of Roy Ascott).

A Semiotic Perspective

A semiotic approach has been taken by design writers, both at the linguistic level, looking at the way objects can be "written" and "read" as visual signs, and at the more general level of the study of consumerism, where semiological analysis of objects as commodities has revealed their part in maintaining what Roland Barthes (1989) has called "mythologies." An impressive semiotic analysis of the object is Baudrillard's (1981) *For a Critique of the Political Economy of the Sign,* which shifts the emphasis on the analysis of commodities away from the production of objects to the consumption of signs. But, as Daniel Miller (1987) writes: "While the rise of semiotics in the 1960s was advantages [sic] in that it provided for the extension of linguistic research into other domains, any of which could be treated as a semiotic system, this extension took place at the expense of subordinating the object qualities of things to their word-like properties" (95-96).

A Material Culture Perspective

Although there is very little available on the electronic object, the study of material culture is still of interest because it situates the object firmly within everyday life. Academically, it is somewhere between anthropology, sociology, and ethnology.

Miller (1987) claims there is a an "extraordinary lack of academic discussion pertaining to artifacts as objects, despite their pervasive presence as the context for modern life" (85), and provides an alternative to the semiological analysis of mass consumption by distinguishing material culture from language and the study of meaning in order to focus on the physical nature of artifacts.

In contrast to analyses of the object in relation to consumerism, Mihaly Csikszentmihalyi and Eugene Roshberg-Halton's (1981) *The Meaning of Things* analyzes the meaning of objects in domestic settings, emphasizing their symbolic role. And in "The Metafunctional and Dysfunctional System: Gadgets and Robots," Baudrillard (1996) writes about the electronic gadget as the subject of a science of imaginary technical solutions. Although originally written nearly thirty years earlier, Baudrillard's analysis of electronic gadgets is far more stimulating than a more recent analysis, *Consuming Technologies,* by Roger Silverstone and Eric Hirsch (1992), which is more concerned with descriptive models than with Baudrillard's challenges to the imagination. *Hertzian Tales* is more concerned with "critical" theories, and thus in assessing the development of objects not against whether they fit into how things are now, but the desirability of the changes they encourage.

The value of material culture for this study is that it draws attention to the complex nature of our relationship to ordinary objects and provides standards against which new electronic objects can be compared.

A Design Perspective

Since the early 1960s a very narrow form of semiotic analysis has dominated design thinking about the electronic object. Of books written about design from a theoretical point of view, only John Thackara's (1988) *Design after Modernism* contains new perspectives on the electronic object.

Books and articles by designers, based on particular projects prove more interesting. Manzini and Susani (1995) present a collection of design projects that explore a place for solidity within the fluid world created by electronic technology: "In the fluid world the permanent features we need are no longer there as a matter of course, but are the result of our desire; the 'solid side' in a fluid world, if and when it exists, will be the result of a design" (16).

Their strong emphasis on aesthetics and ecological concerns is a powerful example of design research carried out by practicing designers within an intellectual context. Susani has developed a design perspective that locates the electronic object within material culture rather than semiology or electronic media. He writes: "We are lacking a discipline, perhaps an 'objectology,' or an 'object ethology,' which allows us to analyse and systematise objects and to formulate the rules and codes of their behaviour... a discipline which recovers and updates the interrupted discourse of material culture, in crisis since the world of objects was taken over by the world of products and the world of consumption" (Susani 1992, 42). He also recommends a sensual approach to introducing technology into the home, building on what is already there, and exploring the overlap between the material and immaterial world from an aesthetic and anthropological point of view. He suggests that material culture could offer useful insights to this problem.

A Literary Perspective

However, the most fruitful reflection on material culture is to be found, not in anthropology or sociology, but in literature concerned with the poetry of everyday objects. In *The Poetics of Space,* Gaston Bachelard (1969) offers an analysis, influenced by psychoanalysis, that emphasises the poetic dimension of humble furniture such as wardrobes and chests of drawers; Jun'ichiro Tanizaki's (1991) *In Praise of Shadows* considers the Japanese object in relation to shadows and darkness, and the effects *of* electricity on their appreciation; and Nicholson Baker's novels (such as *The Mezzanine* (1986) and *Room Temperature* (1990)) give everyday industrial products significant roles.

The view of objects suggested by literary writers reveals a poetry of material culture that offers a fresh alternative to the formal aesthetic criticism of the art object and to the academic analysis of their meaning as signs. Their objects are firmly grounded in everyday life.

The best writing in this area blends anthropology, sociology, and semiology to explore the irrational dimensions of the material culture of everyday life. As the electronic object rarely features in this literature, the discussion in the rest of this book is based mainly on design proposals.

The Electronic as Object

This section discusses four design approaches to the electronic object: packages, fusions, dematerialization, and juxtaposition. They differ in how each addresses the conflict between the solidity of the object and the fluidity of electronic media. Design is viewed here as a strategy for linking the immaterial and the material.

Packages

Commercial design's approach to the electronic object has been to treat it as a package for electronic technology. An example of this, where the aesthetic and conceptual possibilities of the package are thoroughly exploited, is Daniel Weil's *Radio in a Bag* [...], which takes the idea of the designer's role as a packager of technology to the extreme. On one level the electronics provide decoration, while on another, their exposure signals a nonchalance toward technology. The radio's literal flexibility expresses the flexible structural relationship between electronic components, and its transparency attempts to demystify the electronic object. It shows that by taking a playful approach to package design and liberating it from product semantics, even the packaging of electronics can yield interesting results. Ironically, part of the critical success of this design, despite being a package, is its treatment as a thing rather than an image.

Fusion

> The logic of computers is expressed in forces that are averages of the behaviour of many electrons. No machine has ever been so far removed from the world of human experience: the largest aircraft carriers are still infinitely closer to the human scale than the simplest, slowest microcomputers.
>
> —D. J. BOLTER, *Turing's MAN*

The electronic object is a confusion of conceptual models, symbolic logic, algorithms, software, electrons, and matter. The gap between the scales of electrons and objects is most difficult to grasp.

The architect Neil Denari has spoken of the need for the "overcoming of the symbolic," and his view is that architecture must make a connection between the worlds of electromagnetism and spatial inhabitation. But there is greater chance of bridging the gap between electromagnetism and inhabitable space if one where to explore this route through the design of objects rather than buildings.

The first transistor [. . .] is a test-rig for a key electronic component created by inventors who work at the level of both electrons and matter. They organize matter as interacting volumes of electrons, and they offer a possibility for reconciling the scales that separate the worlds of electrons and space. But once these prototype elements have been subjected to the extreme rationalization required by mass production, they become reduced to abstract ultraminiaturized electronic components. Their modernist poetry, based on truth to materials, is lost.

Closing the gap between the scales of electronics and objects by directly manipulating materials as volumes of electrons is a difficult route for designers. This task is essentially limited to scientists, and even their test-rigs will eventually become miniaturized components. *Clock* by Daniel Weil [. . .] captures some of this quality—partly a reaction against miniaturization, its size is based on the largest circuit boards available in the early 1980s. The circuit is composed visually and the wires linking the two main components are made from dining forks. Familiar objects are put into new but natural relationships based on electrical properties.

This approach resembles the way electricity was dealt with in early natural philosophy books that explained electricity in delightfully poetic ways, drawing attention to unusual but real phenomena: "The simultaneous development of both kinds of electricity is illustrated by the following experiment:—Two persons stand on stools with glass legs, and one of them strikes the other with a catskin. Both of them are now found to be electrified, the striker positively, and the person struck negatively, and from both of them sparks may be drawn by presenting the knuckle" (Everett, *Deschanel's Natural Philosophy,* Part 3,4).

The development of "smart materials" is another area where the gap between the electronic and material is being closed, although primarily for technical reasons. Scientists and engineers are developing new materials,designed at a molecular level, that are responsive, dynamic, and almost biological. Although most of these materials are still experimental some, such as electroluminescent laminates and piezoelectric films have been around for several decades.

Manzini (1986) explores the implications of designing with these new smart materials: "The design of this skin, and therefore of the objects that are made with it, is chiefly the design of interactivity with the environment—a scenario for which we must prepare the stage, the sets, and the actors. Imagining the nature of these 'individual objects' is another new chapter in the history of design" (204). Most of Manzini's specially commissioned examples illustrate the miniaturization arising from integrating previously separate mechanisms and their novel decorative possibilities. However, they do not demonstrate the radical aesthetic potential of these materials to open new channels of communication with the environment of electronic objects.

Only Alberto Meda and Denis Santachiara's *Stroke Lamp* (see Manzini 1986) hints at the new relationships between people and machines made possible through new reactive materials. It is controlled by stroking the surface, which is made from an insulating plastic with a copper circuit deposited on it by a photochemical process similar to that used for printed circuits. Although low-tech, it suggests a sensual and playful interaction with everyday objects that might be extended to more complex interactions as more sophisticated materials become available. Andrea Branzi's *Leaf Electroluminescent Lamp* (1988) for Memphis is another application of advanced electrochemical materials for cultural rather than functional innovation [...].

But generally, designers have not exploited the aesthetic dimension of new materials with the same energy that engineers have exploited their functional possibilities (to backlight LCD screens in laptop computers reducing their bulk and weight, e.g., or to illuminate escape routes in aircraft so they can be seen through smoke).

Most work in this area does not encourage poetic and cultural possibilities to converge with practical and technical ones. The outcome is a stream of unimaginative proposals. For example, AT&T has applied for a patent for a coating of colored polymer sandwiched between two thin layers of indium tin oxide that changes color when a low voltage is fed through it; the company plans to use it to enable phones to change color instead of ringing.

Although combinations of matter and information might eventually lead to interactive surfaces, giving rise to new channels of communication between people and an increasingly intelligent artificial environment of objects, most smart materials are still under development, are expensive, and use large amounts of energy to operate. The most interesting materials are not available for design experiments, and one must either use simulations or work with widely available but less sophisticated materials to create emblems of what might be.

Dematerialization

The electronic object is an object on the threshold of materiality. Although "dematerialization" has become a common expression in relation to electronic technology, it is difficult to define in relation to the tangle of logic, matter, and electrons that is the electronic object.

The CPU of an electronic object is, essentially, physically embodied symbolic logic or mathematics. Its "material" representation is the circuit and the components it connects. Symbolic logic describes the workings of the "machine" the object becomes when the program runs. The algorithm is the logical idea behind the program, a strategy that allows symbolic logic to be translated into a programming language (such as C++) and run through the machine, controlling the flows of electrons through its circuitry.

Dematerialization, therefore, means different things depending on what it is defined in relation to: immaterial/material, invisible/visible, energy/matter, software/hardware, virtual/real. But the physical can never be completely dismissed: "Every symphony has its compact disc; every audio experience its loudspeaker; every visual image its camera and video disc. Behind every outward image or symbol lies mechanical support, and if the immateriality of these images and symbols gives rise to a new approach to the relationship between human being and object, the analysis will be one of the individual's connection with the material support underlying the new culture of immateriality" (Moles 1995, 274).

One argument, put forward in the 1980s by the design group Kunstflug, is that values and functions can completely shift from hardware to software, from three to two dimensions, and ultimately to "design without an object." It sounds like an untenable and an oversimple critique of materialism, but during the mid-1980s it drew attention to their ideas. They argued for a change in the attitude to the consumption of objects, calling on industry to produce solutions, not commodities. "Design without an object" could, as part of a cultural movement, offer an alternative to abstinence from consumption while encouraging "the forsaking of things as objects of desire and covetousness."

In the exhibition Design Today, held at the German Museum of Architecture in 1988, Kunstflug offered two examples of this approach: design proposals for *The Electronic Room: Programmable Appearances—Surfaces, Appliances, Comfort* [...] and *Electronic Hand Calculator* [...]. While the room seems only to reinforce stereotypical approaches to the impact of electronics on architectural spaces, the electronic hand calculator became an icon for "design without an object," defining one extreme position in the debate about the impact of electronic technology on objects.

This interest in dematerializing the object for social and political reasons is echoed by the "info-eco" ideas of Manzini, Susani, and Thackara who argue that, by focusing on experiences rather than objects, electronic technology can provide services currently offered through discrete products. In the Info-eco Workshops held at the Netherlands Design Institute in 1995, participants developed scenarios on themes such as "Beyond Being There." Dematerialization was used to investigate hypothetical situations in limited scenarios and discover how information technology might satisfy needs normally fulfilled materially. For instance, telematic tools were proposed where the quality of experience they offered would reduce the desire to travel—digital information being easier to move than matter. (Reports detailing the results of the workshops are available at www.design-inst.nl/.)

In the introduction to the 1994 Ars Electronica festival in Linz, Peter Weibel describes a another form of dematerialization, "intelligent ambience." It arises from shifting emphasis from the "machine" to its "intelligence," and distributing that intelligence throughout an environment:

> Machine intelligence will serve to make the environment more efficient and more intelligent so that it will be able to respond more dynamically and interactively to human beings. The realisation of the concepts of computer aided design and virtual reality will thus be followed by the realisation of computer aided environment and intelligent, interactive, real surroundings. The latter will be referred to as intelligent ambience—an environment based on machine intelligence. One could say: from Tron house to the Tron ambience. (Weibel 1994, n.p.)

Weibel's observations fall between two other views of dematerialization. The first, which belongs to the human factors world, has been referred to as "ubiquitous computing" and is the subject of much research. Dematerialization is seen as a way of providing "transparent" interfaces for computers by embedding the technology in familiar objects and environments and introducing a high degree of automation. At the other extreme is Design Primario, where design effort shifts from hardware to software, and controls levels of light, sound, and temperature to provide sensual environmental qualities. But the aesthetic possibilities of this form of dematerialization have been best exploited by architects: Toyo Ito's design for his *Dreams Room* at the Victoria and Albert Museum in London, was partly motivated by a desire to extend this approach to include information (which he referred to as "active air").

Another form of dematerialization is defined by electronic objects' role as interfaces. With these objects the interface is everything. The behavior of video recorders, televisions, telephones, and faxes is more important than their appearance and physical form. Here design centers on the dialogue between people and machines. The object is experienced as an interface, a zone of transaction. Although most work in this area tends to reduce the object to a "graphical user interface," a screen, designers are beginning to explore the full potential of the "thingness" of the object. The product becomes virtualized and is represented by a set of physical icons and their various permutations. This could lead to more sensual interfaces than screens and offer new aesthetic qualities.

The work of Durrell Bishop offers a vision of what this might mean: existing objects are used as physical icons, material representations of data that refer to both the pragmatic and poetic dimensions of the data being manipulated. The objects and the electronic structure need have nothing in common. For example, in his design for a telephone answering machine, small balls are released each time a message is left. These balls are representations of the pieces of information left in the machine, allowing direct interaction between the owner and the many possibilities an answerphone offers for connecting to telephone and computer systems. If the caller leaves a number, the ball can automatically dial it; if the message is for somebody else, the ball can be placed in his or her personal tray. Although applied very practically, Bishop's thinking engages with the cultural context in which the technology is used. An "aesthetics of use" emerges.

The material culture of non-electronic objects is a useful measure of what the electronic object must achieve to be worthwhile but it is important to avoid merely superimposing the familiar physical world onto a new electronic situation, delaying the possibility of new culture through a desperate desire to make it comprehensible.

Juxtaposition

How can we discover analogue complexity in digital phenomena without abandoning the rich culture of the physical, or superimposing the known and comfortable onto the new and alien? Whereas dematerialization sees the electronic integrated into existing objects, bodies, and buildings, the juxtaposition of material and electronic culture makes no attempt to reconcile the two: it accepts that the relationship is arbitrary, and that each element is developed in relation to its own potential. The physical is as it always has been. The electronic, on the other hand, is regarded only in terms of its new functional and aesthetic possibilities; its supporting hardware plays no significant part.

Fiona Raby's telematic *Balcony* [. . .] demonstrates how the contradictory natures of electronic and material cultures can coexist. The balcony provides access to an open telephone line linking two or three places. Its physical form provides a focal point and support for leaning on, while an ultrasonic sensor detects the approach of users and slowly clears the line. There is no point trying to integrate the physical support and the ultrasonic field, to collapse one into the other, forcing the physical to represent the electronic or to disappear completely so that only electronic effects remain. Juxtaposition allows the best qualities of both to coexist, each with its own aesthetic and functional potential. Technology can be mass-produced whereas the object can be batch-produced. No effort need be made to reconcile the different scales of the electronic and the material. They can simply coexist in one object. They can grow obsolete at different rates as well. Robert Rauschenbergs *Oracle* [. . .] has had its technology updated three times over thirty years, but its materiality and cultural meaning remain unchanged. Cultural obsolescence need not occur at the same rate as technological obsolescence.

Perhaps the "object" can locate the electronic in the social and cultural context of everyday life. It could link the richness of material culture with the new functional and expressive qualities of electronic technology.

In Philips's 1996 *Vision of the Future* project (Philips Corporate Design 1996), a more subtle awareness of the value of material culture has entered the mainstream of design thinking and may well soon enter the marketplace and everyday life. The project consists of over one hundred design proposals for products for five to ten years in the future. But this awareness is primarily expressed in this project by references to existing object typologies—for example, hi-tech medical kits in the form of medicine cabinets—rather than by radically new hybrids. The designers focus more on practical needs, the electronic qualities are not fully exploited, and the types of objects proposed are already familiar from student degree shows. But the designs do achieve a new visual language, sensual, warm, and friendly. They are well-mannered and socially competent. In these projects the electronic object has reached an optimal level of semiotic and functional performance.

The Electronic as Post-optimal Object

The most difficult challenges for designers of electronic objects now lie not in technical and semiotic functionality, where optimal levels of performance are already attainable, but in the realms of metaphysics, poetry, and aesthetics, where little research has been carried out:

> This is what differentiates the 1980s from 1890, 1909, and even 1949—the ability of industrial design and manufacturers to deliver goods that cannot be bettered, however much money you possess. The rich find their exclusivity continuously under threat. . . .
>
> Beyond a certain, relatively low price (low compared with other times in history) the rich cannot buy a better camera, home computer, tea kettle, television or video recorder than you or I. What they can do, and what sophisticated retailers do, is add unnecessary "stuff" to the object. You can have your camera gold plated. (Dormer 1990, 124)

The position of this book is that design research should explore a new role for the electronic object, one that facilitates more poetic modes of habitation: a form of social research to integrate aesthetic experience with everyday life through "conceptual products."

In a world where practicality and functionality can be taken for granted, the aesthetics of the post-optimal object could provide new experiences of everyday life, new poetic dimensions.

Typesetting: Lalit Joshi
Design: Nadine Rinderer
Design Concept BIRD: Christian Riis Ruggaber, Formal
Editors: Simon Grand, Wolfgang Jonas
Copy editing: Susan James
Project Management: Robert Steiger, Sarah Schwarz, Katharina Kulke
Typefaces: Arnhem
Proofreading: Joshua Korn

Graphic Design & Book Production:
ActarBirkhäuserPro www.actarbirkhauserpro.com Barcelona - Basel

A CIP catalogue record for this book is available from the Library of Congress, Washington D.C., USA.

Bibliographic information published by the German National Library. The German National Library lists this publication in the Deutsche Nationalbibliografie; detailed bibliographic data are available on the Internet at http://dnb.d-nb.de.

Distribution:
ActarBirkhäuserD
www.actarbirkhauser-d.com Barcelona - Basel - New York

Roca i Batlle 2
E-08023 Barcelona
T +34 93 417 49 43
F +34 93 418 67 07
salesbarcelona@actarbirkhauser.com

Viaduktstrasse 42
CH-4051 Basel
T +41 61 5689 800
F +41 61 5689 899
salesbasel@actarbirkhauser.com

151 Grand Street, 5th floor
New York, NY 10013, US A
T +1 212 966 2207
F +1 212 966 2214
salesnewyork@actarbirkhauser.com

Printed on acid-free paper produced from chlorine-free pulp. TCF ∞

Printed in Spain

ISBN 978-3-0346-0716-2

9 8 7 6 5 4 3 2 1

www.birkhauser.com